Boron Nitride Nanostructures

Boron Nitride Nanostructures

Special Issue Editors

Philippe Miele
Mikhael Bechelany

MDPI • Basel • Beijing • Wuhan • Barcelona • Belgrade

MDPI

Special Issue Editors

Philippe Miele
Institut Europeen des Membranes
France

Mikhael Bechelany
Institut Europeen des Membranes
France

Editorial Office
MDPI
St. Alban-Anlage 66
4052 Basel, Switzerland

This is a reprint of articles from the Special Issue published online in the open access journal *Nanomaterials* (ISSN 2079-4991) in 2018 (available at: https://www.mdpi.com/journal/nanomaterials/special_issues/boron_nitride_nano#)

For citation purposes, cite each article independently as indicated on the article page online and as indicated below:

LastName, A.A.; LastName, B.B.; LastName, C.C. Article Title. *Journal Name* **Year**, *Article Number*, Page Range.

ISBN 978-3-03897-489-5 (Pbk)
ISBN 978-3-03897-490-1 (PDF)

Cover image courtesy of Philippe Miele and Mikhael Bechelany.

Contents

About the Special Issue Editors

Philippe Miele (Prof. Dr.) received his PhD in Inorganic Chemistry in 1993 from the University of Montpellier. Following a postdoctoral fellowship at Georgia Institute of Technology (Atlanta, GA, USA), he became Assistant Professor (1994), and then Professor at the University Lyon 1. He was the leader of the group "Molecular Precursors and Inorganic Materials" in the Laboratory of Multimaterials and Interfaces (UMR UCBL/CNRS 5615). In 2003, he was appointed to the position of Lab Head up to 2010. In fall 2010, he joined the European Institute of Membranes of Montpellier (UMR CNRS 5635) with a part of his former group setting up the "Molecular Materials and Ceramics" group. In 2011, he was appointed to his present position as Lab Head. His main research interests lie in boron chemistry and in non-oxide advanced ceramics using the Polymer Derived Ceramics route, particularly new boron- and silicon-based materials. He was nominated junior member (2003) then senior member (2016) of the "Institut Universitaire de France" (IUF) and was elected in 2011 to the "World Academy of Ceramics" affiliated with the class "Science".

Mikhael Bechelany (Dr.) obtained his PhD in Materials Chemistry from the University of Lyon (France) in 2006. His PhD work was devoted to the synthesis and characterization of silicon and boron-based 1D nanostructures (nanotubes, nanowires, and nanocables). Then, he worked as a post-doc at EMPA (Switzerland). His research included the fabrication of nanomaterials (nanoparticles and nanowires), their organization, and their nanomanipulation for applications in different fields such as photovoltaic, robotic, chemical, and bio-sensing. In 2010, he became a scientist at CNRS. His current research interest in the European Institute of Membranes of Montpellier (UMR CNRS 5635) in Montpellier (France) focuses on novel synthesis methods for metals and ceramics nanomaterials like Atomic Layer Deposition (ALD), electrodeposition, electrospinning, and/or on the nanostructuring using natural lithography (nanospheres and/or membranes). His research efforts include the design of nanostructured membranes for health, the environment, and renewable´energy.

Preface to "Boron Nitride Nanostructures"

Boron nitride (BN) is a III–V material, well known for its outstanding physico-chemical properties, such as high chemical and thermal stabilities and unique electronic and optical properties. In the past few decades, Boron Nitride nanostructures, such as nanosheets, nanotubes, porous materials, nanocapsules, etc., have attracted a great deal of interest because of their potential applications in functional devices.

The research topic of this Special Issue will consider (i) the design of nanostructured boron nitride nanostructures with controlled crystal structures, porosity, and dimensionality, (ii) the functionalization of boron nitride, and (iii) prospective applications of boron nitride nanostructures and materials. It contains six papers dealing with (i) the exfoliation of hexagonal Boron Nitride (h-BN) in liquid phase by ion intercalation, (ii) effective mechanical properties and thickness determinations of Boron Nitride nanosheets using molecular dynamics simulation, (iii) direct observation of inner-layer inward contractions of multiwalled Boron Nitride nanotubes upon in situ heating, (iv) the alignment of Boron Nitride nanofibers in epoxy composite films for thermal conductivity and dielectric breakdown strength improvement, (v) the effect of Boron Nitride on the thermal and mechanical properties of poly(3-hydroxybutyrate-co-3-hydroxyvalerate), and (vi) hexagonal Boron Nitride functionalized with Au nanoparticles—properties and potential biological applications.

Philippe Miele, Mikhael Bechelany
Special Issue Editors

nanomaterials

MDPI

Article

Exfoliation of Hexagonal Boron Nitride (h-BN) in Liquide Phase by Ion Intercalation

Danae Gonzalez Ortiz [1], Celine Pochat-Bohatier [1], Julien Cambedouzou [2], Mikhael Bechelany [1,*] and Philippe Miele [1,3,*]

[1] Institut Européen des Membranes (IEM), UMR-5635, CNRS, ENCSM, University of Montpellier, Place Eugene Bataillon, 34095 Montpellier, France; danae.gonzales-ortiz@umontpellier.fr (D.G.O.); celine.pochat@umontpellier.fr (C.P.-B.)
[2] Institut de Chimie Séparative de Marcoule (ICSM), CEA, CNRS, ENSCM, University of Montpellier, 30207 Marcoule, France; julien.cambedouzou@enscm.fr
[3] Institut Universitaire de France (IUF), 1 Rue Descartes, 75231 Paris CEDEX 5, France
* Correspondences: Mikhael.bechelany@umontpellier.fr (M.B.); Philippe.miele@umontpellier.fr (P.M.); Tel.: +33-4-6714-9167 (M.B.)

Received: 17 July 2018; Accepted: 5 September 2018; Published: 12 September 2018

Abstract: A green approach to prepare exfoliated hexagonal boron nitride nanosheets (h-BNNS) from commercially pristine h-BN involving a two-step procedure was investigated. The first step involves the dispersion of pristine h-BN within an aqueous solution containing gelatin and potassium or zinc chloride using a sonication method. The second involves the removal of larger exfoliated h-BNNS through a centrifugation procedure. The exfoliation was caused not only by the sonication effect but also by intercalation of K^+ and Zn^{2+} ions. Transmission electronic microscopy, X-ray diffraction and Raman spectroscopy techniques show that the obtained h-BNNS generally display a thickness of about a few (2–3) layers with an exfoliation efficiency as high as $16.3 \pm 0.4\%$.

Keywords: liquid exfoliation; boron nitride nanosheets; ion intercalation; gelatin

1. Introduction

Two-dimensional (2D) nanosheets have been receiving great attention in recent years, due to their ultrathin structure with large planar dimensions and their outstanding properties [1–3]. The advantage of these materials is that they have great potential in a huge range of applications at scientific and technological levels due to their exfoliated state, as single- or few-layers. The most widely studied 2D layered material is graphene. It is ideally composed of a one-atom thick planar sheet of sp^2 hybridized carbon atoms. Recently, other 2D layered materials, such as layered transition metal dichalcogenides (LTMDs) (e.g., MoS_2 and WS_2), metal oxides and hexagonal boron nitride (h-BN) have gained a renewed interest. Their lamellar structure can be subjected to exfoliation leading to single-layer and few-layers nanosheets. The structure and morphology of exfoliated 2D-layered materials strongly influence their properties; they present high surface area and quantum confinement effects [4,5] and also high chemical stability as well as insulating properties [6].

In this study, we focus our attention on h-BN. Its layered structure consists of a lattice similar to graphite in which boron (B) and nitrogen (N) atoms alternate in a hexagonal arrangement within the layer whereas they are in an eclipsed configuration when observed perpendicularly. Therefore, boron nitride nanosheets (BNNSs) are basically composed of sheets of sp^2 hybridized 2D layers, organized in honeycomb geometry, with an interlayer distance of ca. 0.33–0.34 nm. Recently, the interest of scientists in BNNSs as a 2D nanomaterial has increased because of their advantages compared to graphene. Besides the unique chemical and thermal stability of BNNS [7,8], they also present great mechanical strength and electrical insulating properties (bandgap of 5~6 eV) [9]. Therefore, BNNSs have important applications in nanodevices or functional composites.

The aim of this study is to develop a new exfoliation method to obtain mono-layered or few-layered BNNS, minimizing the use of organic solvents and yielding a large amount of material. Nowadays, two different approaches exist for large-scale production of BNNS, bottom-up and top-down processes. The bottom-up approach includes chemical vapor deposition [10–12] and segregation methods [13]. This methodology provides single-layer BNNS but requires extreme conditions of temperature and pressure. Therefore, it is widely considered as unsuitable to synthesize BNNS on larger scales. On the other hand, the top-down approaches are based on exfoliating bulk h-BN crystals via mechanical [14–16] or sonication methods [17–19]. Such approaches have already been reported for graphene, for which individual sheets of 2D crystals were successfully separated. However, the B–N bond presents a partially ionic character compared to the covalent C–C bonding of graphene, leading to interactions between adjacent concentric shells or neighboring BN layers (so-called "lip-lip" interaction) [7].

Liquid-phase exfoliation is also one of the most investigated methods for exfoliating h-BN. Different attempts to prepare h-BNNS have been reported, such as treating bulk h-BN in various organic solvents. Zhi et al. dissolved h-BN in DMF applying vigorous ultrasonication followed by exfoliation. The exfoliation is attributed in this case to the strong interactions between DMF and the BN surface [18]. Warner et al. used 1,2-dicholoethane to exfoliate bulk h-BN by ultrasonic bath [19]. Zhou et al. demonstrated that they can obtain a highly stable suspension of exfoliated BN using a mixture of solvents such as water and ethanol [20]. Coleman et al. proposed a number of different solvents based on Hansen solubility to exfoliate h-BN. They found the best results with N-methylpyrrolidone (NMP) and isopropanol (IPA) [21]. Ye et al. exfoliated bulk h-BN powder in chloroform under sonication using hyperbranched polyethylene (HBPE) as stabilizer. They obtained stable dispersion of h-BNNS in chloroform [22]. Griffin et al. used an aqueous solution of sodium chlorate combined with ultrasonic power to exfoliate h-BN. The resultant solution was subjected to a centrifugation cascade to isolate smaller nanosheets [23]. Furthermore, molten hydroxides have been used to exfoliate h-BN. Sodium hydroxide (NaOH), potassium hydroxide (KOH) and BN were ground, then they were transferred into an autoclave and heated up to 180 °C. BNNS were obtained as a result [24]. Using these methods, a small yield of around 5% could be achieved. Besides organic solvents, water has been used as a solvent for exfoliation of h-BN using ultrasounds. Gelatin-assisted h-BN exfoliation under ultrasonic conditions has been also reported without large success concerning the exfoliation yield [25]. All these approaches failed to produce h-BNNS with high-yield exfoliation and in an environmentally friendly way. Therefore, a new low-cost method to produce large scale and high-quality h-BNNS needs to be explored.

Herein, we present a facile and green procedure combining sonication and centrifugation methods to produce high-yield h-BNNS from a commercially available powder. Gelatin and chloride salts were used to assist the exfoliation. The ion intercalation between the layered structures of bulk material facilitates the exfoliation. The obtained h-BNNS generally displays a thickness of about a few (3–9) layers with an exfoliation efficiency as high as 16.3%, and has high crystallinity.

2. Materials and Methods

2.1. Materials

Hexagonal boron nitride (h-BN) was supplied from Saint Gobain (Cavaillon, France) (CAS No. 10043-11-5, 95% purity, 325 mesh, 3 μm particle size). Zinc chloride ($ZnCl_2$, 99.99% trace metal basis), potassium chloride (KCl, ≥99%) and gelatin type A (gel strength 300) were purchased from Sigma Aldrich (St. Quentin Fallavier, France). In all the experiments deionized pure water (18 MΩ) was used.

2.2. Exfoliation of h-BN

Hexagonal boron nitride (h-BN) nanosheets were fabricated using liquid-phase exfoliation with the assistance of an ultrasound device (BANDELIN, Berlin, Germany) (model SONOPLUS HD 3100, 100 W, 20 kHz) with a 3 mm diameter microtip (MS73). 1.0 g of pristine h-BN was added to 100 mL of water. The solution was heated up to 80 °C, then 20 g of gelatin were added. The mixture was kept under stirring at the same temperature until complete dissolution of the gelatin. Different concentrations of $ZnCl_2$ and KCl were added to the mixture. The dispersion was kept in a bath at 50 °C to avoid the gelatin solidification and it was sonicated for 3 h at 65% amplitude with pulse off/on 0.5–1 s. After sonication, the yellowish suspension was subjected to two centrifugation steps. In a first step, the solution was centrifuged at 3000 rpm for 30 min. This first centrifugation step ensures the separation of large h-BNNS from light ones by remaining at the bottom of the solution. Then, the supernatant was collected and subjected to a second centrifugation step at 6000 rpm for 30 min. The supernatant was collected and dried at 60 °C overnight. In a first step, the resultant material was heated up to 600 °C under air to remove the gelatin. In a second step, the obtained powders were heated up to 1000 °C under Ar atmosphere with a flux of 200 mL/min, to improve the h-BNNS crystallinity. The obtained h-BNNS were washed several times with water and ethanol to remove any remaining impurities.

2.3. Characterization Techniques

The transmission electron microscopy (TEM) was performed using a JEOL 2200 FS-200 kV equipped with a STEM (Scanning Transmission Electron Microscopy, JEOL, Tokyo, Japon) module a Bright Field (BF) detector and a CCD Gatan USC 4092x4092px^2 camera. To carry out the analysis, a drop of h-BNNS/isopropanol dispersion was deposited on the surface of the carbon grids. Atomic force microscopy (AFM) analysis was performed using an AFM NANOMAN 5 from Veeco Instrument controlled with the Nanoscope V software V8 (Bruker Corporation, Coventry, UK). The exfoliated h-BN was previously dissolved in water and then a drop of the solution was deposited on a silicon wafer. X-ray diffraction patterns of exfoliated h-BNNS were recorded using a PANalytical Xpert powder X-ray diffraction (XRD) system (PANalytical, Almelo, Hollande) with Cu Kα radiation, a scan speed of 2° min^{-1}, a 2θ range between 3° and 70°, and a step rate about 0.02° per second. The Fourier transformed infrared (FTIR) spectra were recorded with a NEXUS instrument (Thermo Fisher Scientific, Waltham, USA) equipped with an attenuated total reflection (ATR) accessory in the frequency range of 600–4000 cm^{-1}. Raman spectra have been obtained from a Horiba xplora, λ = 659 nm. The yield of exfoliated h-BNNSs was calculated from the following equation:

$$Yield\ (\%) = (Weight\ (h\text{-}BNNS)/Weight\ (h\text{-}BN)) \times 100 \tag{1}$$

3. Results

The starting h-BN powder has a typical lateral particle size in the range of approximately 3 μm. The preparation of h-BNNS was carried out from h-BN pristine in a simple combination of gelatin/chloride salts sonication and centrifugation steps. Additional heating steps were performed to remove the gelatin and improve the h-BNNS crystallinity. The yield of obtained h-BNNS was calculated from Equation (1). The results are shown in Table 1. The best exfoliation efficiency (16.3%) is obtained when 1.0 wt % of KCl was used to intercalate the ions within the layered structure of h-BN (Table 1).

Table 1. Yields of h-BNNS obtained through exfoliation with gelatin, and gelatin assisted ion intercalations with 0.5 and 1.0 wt % of KCl and $ZnCl_2$.

Sample	Initial h-BN (g)	Purified h-BNNS (g)	Yield (%)
Gelatin	1.015	0.065	6.4 ± 0.1
0.5 wt % K/h-BNNS	1.014	0.143	14.1 ± 0.2
1.0 wt % K/h-BNNS	1.016	0.165	16.3 ± 0.4
0.5 wt % Zn/h-BNNS	1.005	0.124	12.3 ± 0.2
1.0 wt % Zn/h-BNNS	1.009	0.108	10.8 ± 0.2

3.1. Transmisson Electron Microscopy

Transmission electron microscopy (TEM) studies were carried out in order to determine the morphology of h-BNNS. The as-obtained dispersion showed the presence of irregularly shaped h-BNNS with lateral sizes ranging from a few tens of nanometers to as large as over 0.5 μm (Figure 1). The relative thickness of the h-BNNS could be visually examined by the transparency/darkness of the nanosheets against the electron beam, with thinner nanosheets appearing lighter. It is observed that the h-BNNS flakes obtained when the exfoliation is assisted by KCl (Figure 1a,b) display a lateral size mostly less than 200 nm, and the thickness is about few (2–3) layers. White arrows were added to the TEM images to facilitate the identification of the nanosheets. These results indicate that the sheets reduced their size by "cutting" the large h-BN sheets into smaller ones. Furthermore, the sonication process allows an effective "peeled out" of nanosheets from the bulk h-BN material, forming mono-layered and few-layered nanosheets with reduced lateral size in the dispersion, since the initial size of pristine h-BN is about 3 μm.

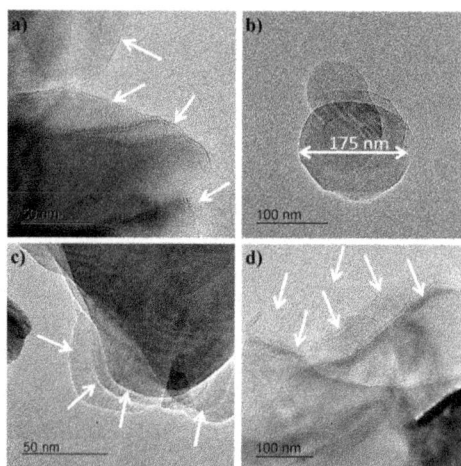

Figure 1. TEM images of h-BNNS exfoliated with (**a**) 0.5 wt % KCl; (**b**) 1.0 wt % KCl; (**c**) 0.5 wt % $ZnCl_2$ and (**d**) 1.0 wt % $ZnCl_2$.

The nanosheets obtained when higher amounts of salts are used to exfoliate the pristine h-BN are thinner than when a lower concentration of salts is used. This could be due to the fact that at higher concentrations, more ions can be intercalated into the layered structure of h-BN, weakening the van der Waals interactions between the layers and leading to greater exfoliation.

In the case of h-BNNS obtained when the exfoliation is assisted by $ZnCl_2$ (Figure 1c,d), the flakes display a lateral size up to 0.5 μm and are composed of a more significant amount of layers. The exfoliation efficiency using $ZnCl_2$ is smaller compared to the exfoliation involving KCl.

These results might be explained regarding ionic diameters. The ionic diameter of K^+ is 2.76 Å while the ionic diameter for Zn^{2+} is 1.48 Å. Taking into account that the interlayer distance of pristine h-BN is 3.3 Å, both ions are able to intercalate between the h-BN layers, but K^+ ions are larger than Zn^{2+} ions. Thus, the incorporation of K^+ ions should have a higher ability to weaken the inter-sheet interactions and to destabilize the packing of BN sheets. This results in a higher exfoliation degree. This approach allows the reduction of the lateral size of h-BNNS as well as a reduction of the number of layers.

3.2. Atomic Force Microscopy

h-BNNS obtained after liquid exfoliation in water assisted by gelatin and by gelatin with $ZnCl_2$ or KCl using sonication tip, were characterized as well by AFM. A number of about 100 h-BNNS have been considered for each sample. The images (Figure 2) indicate a reduction on the thickness of pristine h-BN from 3 μm to 1–3 nm in the case of intercalation of K^+ ions, and reduction of thickness to 2–6 nm for the intercalation of Zn^{2+} ions. Based on these AFM images and taking into account the width of one nanosheet (0.33 nm), we could assume that our h-BNNS are generally composed of 3–9 layers in the case of exfoliation with 1 wt % KCl. Besides, the h-BNNS exfoliated with 1 wt % $ZnCl_2$ displays a thickness of about 6–48 nm, which corresponds to around 2–6 layers. If we compare these results with those when the exfoliation is assisted only with gelatin, we observe that the exfoliation degree is more important when ions assist the exfoliation than when only gelatin is added. The thickness, the number of layers and the lateral size of the nanosheets are summarized in Table 2.

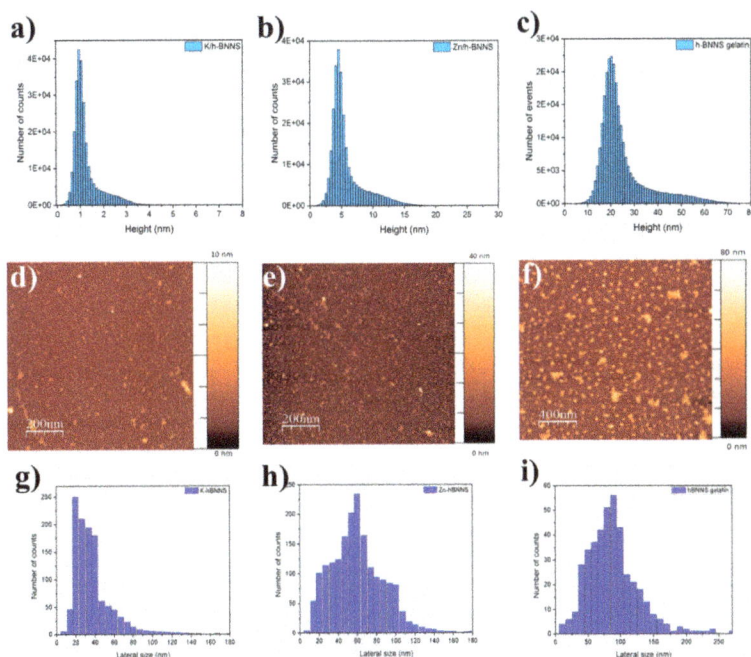

Figure 2. (**a**) Height of h-BNNS exfoliated with 1.0 wt % KCl; (**b**) Height of h-BNNS exfoliated with 1.0 wt % $ZnCl_2$; (**c**) Height of h-BNNS exfoliated with gelatin; (**d**) AFM image of h-BNNS intercalated with 1.0 wt % KCl; (**e**) AFM image of h-BNNS intercalated with 1.0 wt % $ZnCl_2$; (**f**) AFM image of h-BNNS intercalated with gelatin; (**g**) Lateral size of h-BNNS exfoliated with 1.0 wt % KCl; (**h**) of h-BNNS exfoliated with 1.0 wt % $ZnCl_2$ and (**i**) Lateral size of h-BNNS exfoliated with gelatin.

Table 2. Summary of thickness of pristine h-BN and exfoliated h-BNNS.

Sample	Thickness	N° of Layers	Lateral Size (nm)
h-BN	3 μm	-	-
h-BNNS gelatin	10–70 nm	3–200	20–250
1.0 wt % Zn/h-BNNS	2–16 nm	6–48	10–120
1.0 wt % K/h-BNNS	1–3 nm	3–9	10–80

Otherwise, when the exfoliation is assisted only with gelatin, we observed that the nanosheet size is reduced in comparison to pristine h-BN, from 3 μm to about 10–80 nm. If we compare these resulting nanosheets with the ones obtained when the exfoliation is assisted with gelatin/Zn^{2+} or gelatin/K^+ ions, the exfoliation degree is much lower (Figure 2g–i) Furthermore, the h-BNNS obtained through ion intercalation are composed of around 3–9 layers of lateral size about 10–80 nm in case of exfoliation with 1 wt % KCl, while the h-BNNS resulting from gelatin exfoliation are composed of around 30–200 layers of lateral size 20–250 nm. Thus, we can say that the addition of ions to the solution during sonication helps to improve the exfoliation degree and to obtain nanosheets with smaller dimensions.

3.3. X-Ray Diffraction

X-ray diffraction (XRD) was used to investigate the phase structure of the as-prepared materials. h-BN (Figure 3) displays the diffraction peaks at 2θ = 26.66°, ~40–45° and 55.05° which can be correlated to the (002), unresolved (100) and (101), and (004) planes of the hexagonal phase of BN. The unresolved reflections will appear as (10×) in the graphics. No impurity from salts was detected by XRD after h-BNNS were washed (Figure S1), which indicates that the as-prepared h-BNNSs are pure. Energy-dispersive X-ray spectroscopy (EDX) analyses were performed to confirm the purity of obtained h-BNNS (Figure S2) (Tables S1 and S2). However, it is interesting to note that the 002 peak of h-BNNS, exfoliated via intercalation of Zn^{2+} or K^+ ions, becomes broader regarding pristine h-BN.

The full width at half maximum (FWHM) of a peak is sensitive to the variation of the microstructure and stress-strain of the materials. For that, the FWHM of the 002 diffraction peak of h-BNNS was calculated from the XRD patterns. The results are shown in Table 3. It is observed that the FWHM increases when the h-BN is exfoliated with KCl and ZnCl2. The h-BNNS intercalated with 1 wt % of K^+ and Zn^{2+} ions display a higher value of FWHM, indicating that the crystallite size might decrease during the exfoliation process. As is already known in the literature, the peak broadness is related to the crystallite size and it varies inversely with the FWHM: as the crystallite becomes smaller, the peak becomes broader [26]. For this reason, we also calculated the crystallite size of the h-BN pristine, the h-BN exfoliated with gelatin, and the h-BNNSs intercalated with Zn^{2+} and K^+ at different concentrations using the Scherrer equation [27]. The results of the calculations are shown in Table 3.

Table 3. Position of the 002 diffraction peak, *d* spacing and crystallite size of h-BN, h-BNNS-gelatin and h-BNNS intercalated with different concentrations of KCl and ZnCl2.

Sample	Peak Position (°)	*d*-Spacing (nm)	FWHM β (°)	Crystallite Size (nm)
h-BN	26.69	0.334	0.4	20.6
h-BNNS gelatin	26.61	0.335	0.4	20.1
0.5 wt % K/h-BNNS	26.55	0.335	1.6	4.9
1.0 wt % K/h-BNNS	26.55	0.335	2.9	2.9
0.5 wt % Zn/h-BNNS	26.61	0.335	0.7	10.7
1.0 wt % Zn/h-BNNS	26.59	0.335	1.9	4.2

The results obtained by the Scherrer equation show that increasing the salt (ZnCl2 and KCl) concentration during the intercalation results in smaller h-BNNS crystallites as compared to pristine h-BN. In the case of exfoliation assisted with gelatin only, the reduction of crystallite size is less pronounced. The initial crystallite size in pristine h-BN is calculated to be 20.6 nm. It has to be pointed out that the crystallite size calculated from Scherrer equation assumes that all crystallites have the same size and shape, and also that the crystallite size is different than the particle size (a particle may be made up of several different crystallites). This explains the differences emerging from the comparison of the results from XRD and AFM techniques. On one hand, the addition of the K^+ ions during sonication shifts the peak to lower angles, from $2\theta = 26.69°$ in pristine h-BN to $26.55°$ in h-BNNS. On the other hand, it reduces the crystallite size of the resulting h-BNNS down to 4.9 nm when 0.5 wt % KCl was used and down to 2.9 nm when 1.0 wt % KCl was introduced. In the case of intercalation with Zn^{2+} ions, the peak is also shifted to lower angles, from $2\theta = 26.69°$ in pristine h-BN to 26.61 and $26.59°$ in h-BNNS intercalated with 0.5 and 1.0 wt % $ZnCl_2$, respectively.

The crystallite size is also reduced in comparison to pristine h-BN: from 20.6 nm to 10.7 nm and 4.2 nm when 0.5 and 1.0 wt % $ZnCl_2$ were introduced, respectively. In the case of h-BN exfoliation assisted by gelatin, the FWHM is slightly reduced if compared with pristine h-BN, thus crystallite size is also reduced from 20.6 nm to 20.1 nm. This shift of the peak position in the exfoliated h-BNNS is related to the interlayer distance (*d*-spacing). For this reason, the interlayer distance of h-BN and h-BNNS intercalated with 0.5 wt % and 1.0 wt % of KCl and $ZnCl_2$ was calculated using Bragg law. The results are shown in Table 3. It is observed that the intercalation either with K^+ or Zn^{2+} ions has an influence on the BN interlayer distance. Pristine h-BN displays a *d*-spacing of 0.334 nm. The intercalation of 0.5 wt % and 1.0 wt % of KCl gives a larger interlayer distance, i.e., 0.335 nm. Note that the increase of the interlayer distance is too small to be due to remaining ions between the layers. It is more probably due to structural defects coming from the sonication process in the BN planes that modifies their equilibrium distance.

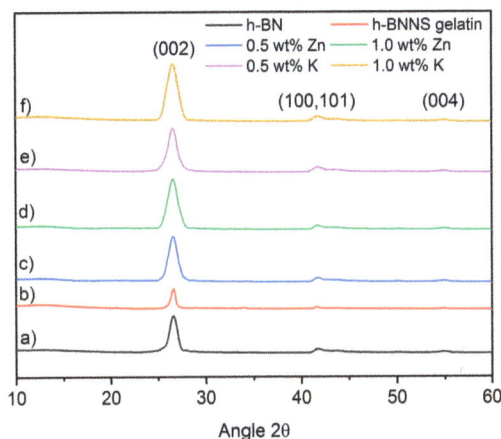

Figure 3. XRD patterns (**a**) pristine h-BN; (**b**) h-BNNS gelatin; (**c**) h-BNNS exfoliated by intercalation of 0.5 wt % $ZnCl_2$; (**d**) h-BNNS exfoliated by intercalation of 1.0 wt % $ZnCl_2$; (**e**) h-BNNS exfoliated by intercalation of 0.5 wt % KCl; and (**f**) h-BNNS exfoliated by intercalation of 1.0 wt % KCl.

On the other hand, the intercalation of Zn^{2+} ions produces the same effect; the interlayer distance is increased to 0.335 nm when 0.5 wt % and 1.0 wt % of $ZnCl_2$ was used, respectively. In the case of the addition of gelatin alone, the interlayer distance also increases up to 0.335 nm but as we mentioned before, the crystallite size is larger than when the exfoliation is assisted with both gelatin and ions.

3.4. Raman Spectroscopy

Raman spectra of pristine h-BN, h-BNNS gelatin and h-BNNS exfoliated with the intercalation of K^+ ions are shown in Figure 4. h-BN and h-BNNS spectra exhibit a characteristic peak due to the typical B–N stretching mode (E_{2g}). Table 4 shows the Raman shifts of pristine h-BN, h-BNNS gelatin and h-BNNS intercalated with 0.5 wt % and 1.0 wt % KCl and ZnCl$_2$. Pristine h-BN displays this peak at ~1366 cm^{-1}. It is observed that the peak intensity becomes progressively weaker and broader in both cases: exfoliation with gelatin and with ions intercalation. In addition, the increase of ion concentration makes this effect more pronounced. The exfoliation assisted with gelatin results in a small shift to higher wavenumbers of this peak at about 1367 cm^{-1}. The exfoliation using 0.5 wt % and 1.0 wt % of KCl results in the shift of the E_{2g} phonon mode to approximately 1370 cm^{-1}, corresponding to a blue shift of ~4 cm^{-1}. In general, the shift in the peak position can be linked to strain conditions within the layers [11,28,29]. Thus, results agree with an effective exfoliation of h-BN into thinner flakes, which leads to a higher in-plane strain and weaker interlayer interactions [30–32]. This effect is more pronounced when increasing the ion concentration [33].

Table 4. Raman shift of h-BN, h-BN gelatin and h-BNNS intercalated with different concentrations of KCl and ZnCl$_2$.

Sample	Raman Shift (cm^{-1})
h-BN	1365.7 ± 0.1
h-BNNS gelatin	1366.8 ± 0.3
0.5 wt % K/h-BNNS	1369.3 ± 0.4
1.0 wt % K/h-BNNS	1370.2 ± 0.2
0.5 wt % Zn/h-BNNS	1368.1 ± 0.2
1.0 wt % Zn/h-BNNS	1368.1 ± 0.3

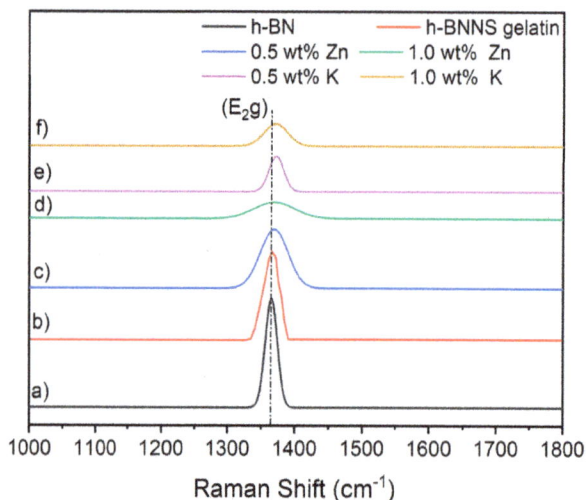

Figure 4. Comparative Raman spectra of (**a**) pristine h-BN; (**b**) h-BNNS gelatin; (**c**) h-BNNS exfoliated by intercalation of 0.5 wt % ZnCl$_2$; (**d**) h-BNNS exfoliated by intercalation of 1.0 wt % ZnCl$_2$; (**e**) h-BNNS exfoliated by intercalation of 0.5 wt % KCl; and (**f**) h-BNNS exfoliated by intercalation of 1.0 wt % KCl.

Raman spectra of h-BNNS exfoliated with the intercalation of Zn^{2+} ions are also shown in Figure 4. Similar to the exfoliation of h-BN with K$^+$, the peak intensity becomes progressively weaker and broader as the ion concentration increases. The E_{2g} phonon mode of 0.5 wt % and 1.0 wt % of ZnCl$_2$ is centered at approximately 1368 cm^{-1}, corresponding to a ~2 cm^{-1} blue shift. In this case, the shift is lower than for the intercalation of K$^+$. This fact is in line with the lower degree of exfoliation that has been evidenced by the others characterization techniques presented above.

3.5. Fourier-Transform Infrared Spectroscopy

Figure 5 shows the FTIR spectra of pristine h-BN and h-BNNS intercalated with K^+ and Zn^{2+} ions. Two strong FTIR bands at ~1380 cm^{-1} and ~812 cm^{-1} are present in pristine h-BN, which are attributed to the B–N stretching (in-plane ring vibration, E_{1u} mode) and the B–N–B bending (out-of-plane vibration, A_{2u} mode), respectively. Table 5 shows the FTIR bands attribution and peak positions presented on the spectra. Intercalated h-BN with 0.5 wt % and 1.0 wt % KCl and $ZnCl_2$ present the same bands than the pristine h-BN.

In addition, the FTIR spectra of h-BNNS intercalated with 0.5 wt % and 1.0 wt % KCl and $ZnCl_2$ show another absorption peak at ~3200 cm^{-1}, which could be ascribed to hydroxyl group (–OH) vibration. This peak might appear due to the large number of defects such as vacancy defects, dislocations and exposed edges introduced on the h-BNNS surfaces during sonication. Also, a new bending mode appears at ~1200 cm^{-1} which is correlated to (–OH) vibration.

Table 5. FTIR bands attribution and peak positions of h-BN and h-BNNS intercalated with different concentrations of KCl and $ZnCl_2$.

Sample	Attribution	Peak Position
h-BN	B–N stretching	1270.9
	B–N–B bending	759.7
0.5 wt % K/h-BNNS	O–H stretching	3372.9
	B–N stretching	1359.6
	B–N–B bending	779.7
1.0 wt % K/h-BNNS	O–H stretching	3340.2
	B–N stretching	1365.4
	B–N–B bending	779.1
0.5 wt % Zn/h-BNNS	O–H stretching	3355.6
	B–N stretching	1336.5
	B–N–B bending	767.5
1.0 wt % Zn/h-BNNS	O–H stretching	3367.2
	B–N stretching	1288.2
	B–N–B bending	742.5

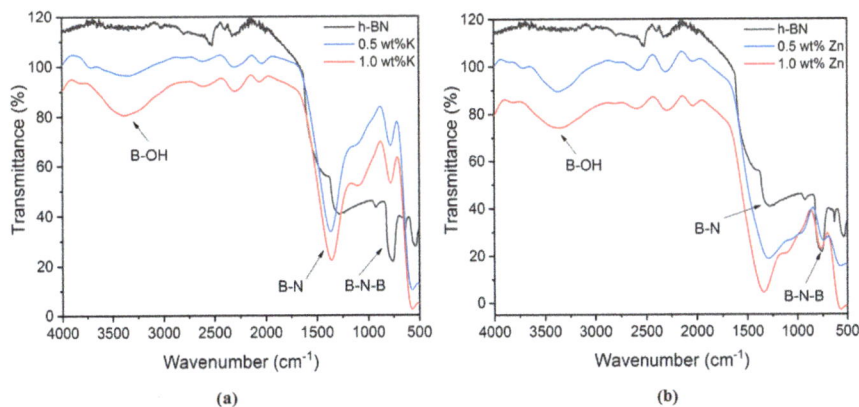

Figure 5. (**a**) FTIR spectra of pristine h-BN of and h-BNNS intercalated with Zn^{2+} at different concentrations (0.5 wt % and 1.0 wt % KCl); and (**b**) FTIR spectra of pristine h-BN of and h-BNNS intercalated with Zn^{2+} at different concentrations (0.5 wt % and 1.0 wt %).

The exfoliation of these materials can be explained by the effect of acoustic cavitation of high-frequency ultrasound in the formation, growth and collapse of microbubbles in solution. This effect induces shock waves on the surface of the bulk material causing exfoliation. In addition, the introduced ions (K^+ and Zn^{2+}) could be inserted into the h-BN layers, inducing an increase in the interlayer spacing. Furthermore, the non-polar chains of the gelatin can adsorb on the surface of flakes through hydrophobic–hydrophobic interactions, which results in the formation of a stable dispersion. Further research concerning reaction conditions, such as varying the reaction time, the applied power or the reaction temperature, might be performed to improve the reaction yield.

4. Conclusions

In summary, a green approach to prepare exfoliated h-BNNS from commercially pristine h-BN involving a two-steps procedure was investigated. The exfoliation was caused not only by sonication effect but was also assisted by gelatin and intercalation of K^+ and Zn^{2+} ions. The latter resulted in the observation of few-layered h-BNNS of reduced lateral size, as well as few-layered h-BNNS with a lateral size below 200 nm, as evidenced from TEM images. The yield for h-BNNS intercalated with 1.0 wt % KCl was about $16.3 \pm 0.4\%$. The crystallinity of the obtained h-BNNS was confirmed using XRD. Raman microscopy further confirmed the presence of the few-layered flakes, by analyzing the peak position shift from the pristine h-BN. The FTIR showed a band corresponding to –OH functions in the obtained h-BNNS through the intercalation of the ions. These functions could be attributed to the functionalization of h-BNNS edges or vacancy defects. Future works are in progress in order to improve the yield of the exfoliated h-BNNS by controlling the different factors that could affect the exfoliation, such as the size of pristine h-BN, the sonication conditions (time, power or pulse) or the centrifugation steps.

Supplementary Materials: The following are available online at http://www.mdpi.com/2079-4991/8/9/716/s1, Figure S1: XRD patterns of exfoliated h-BNNS in presence of $ZnCl_2$ without further washing, Figure S2: (a) EDX image of h-BNNS exfoliated with 1.0 wt % $ZnCl_2$; (b) Element mapping images of the h-BNNS exfoliated with 1.0 wt % $ZnCl_2$, Table S1: EDX descriptive analysis of h-BNNS exfoliated with 0.5 wt % and 1.0 wt % KCl, Table S2: EDX descriptive analysis of h-BNNS exfoliated with 0.5 wt % and 1.0 wt % $ZnCl_2$.

Author Contributions: D.G.O., M.B. and P.M. conceived and designed the experiments; D.G.O. performed the experiments; D.G.O., J.C., C.P.-B., M.B. and P.M. analyzed the data; D.G.O. wrote the paper.

Funding: This research was funded by the French National Agency (ANR) through the LabEx CheMISyst grant number [ANR-10-LABX-05-01].

Conflicts of Interest: The authors declare no conflict of interest.

References

1. Novoselov, K.S.; Geim, A.K.; Morozov, S.V.; Jiang, D.; Zhang, Y.; Dubonos, S.V.; Grigorieva, I.V.; Firsov, A.A. Electric field effect in atomically thin carbon films. *Science* **2004**, *306*, 666–669. [CrossRef] [PubMed]
2. Huang, X.; Qi, X.; Boey, F.; Zhang, H. Graphene-based composites. *Chem. Soc. Rev.* **2012**, *41*, 666–686. [CrossRef] [PubMed]
3. Huang, X.; Yin, Z.; Wu, S.; Qi, X.; He, Q.; Zhang, Q.; Yan, Q.; Boey, F.; Zhang, H. Graphene-based materials: Synthesis, characterization, properties, and applications. *Small* **2011**, *7*, 1876–1902. [CrossRef] [PubMed]
4. Mak, K.F.; Lee, C.; Hone, J.; Shan, J.; Heinz, T.F. Atomically thin MoS_2: A new direct-gap semiconductor. *Phys. Rev. Lett.* **2010**, *105*, 136805. [CrossRef] [PubMed]
5. Nicolosi, V.; Chhowalla, M.; Kanatzidis, M.G.; Strano, M.S.; Coleman, J.N. Liquid exfoliation of layered materials. *Science* **2013**, *340*, 1226419. [CrossRef]
6. Xu, M.; Liang, T.; Shi, M.; Chen, H. Graphene-like two-dimensional materials. *Chem. Rev.* **2013**, *113*, 3766–3798. [CrossRef] [PubMed]
7. Golberg, D.; Bando, Y.; Huang, Y.; Terao, T.; Mitome, M.; Tang, C.; Zhi, C. Boron nitride nanotubes and nanosheets. *ACS Nano* **2010**, *4*, 2979–2993. [CrossRef] [PubMed]
8. Chen, Y.; Zou, J.; Campbell, S.J.; Le Caer, G. Boron nitride nanotubes: Pronounced resistance to oxidation. *Appl. Phys. Lett.* **2004**, *84*, 2430–2432. [CrossRef]

9. Kubota, Y.; Watanabe, K.; Tsuda, O.; Taniguchi, T. Deep ultraviolet light-emitting hexagonal boron nitride synthesized at atmospheric pressure. *Science* **2007**, *317*, 932–934. [CrossRef] [PubMed]

10. Shi, Y.; Hamsen, C.; Jia, X.; Kim, K.K.; Reina, A.; Hofmann, M.; Hsu, A.L.; Zhang, K.; Li, H.; Juang, Z.-Y. Synthesis of few-layer hexagonal boron nitride thin film by chemical vapor deposition. *Nano Lett.* **2010**, *10*, 4134–4139. [CrossRef] [PubMed]

11. Song, L.; Ci, L.; Lu, H.; Sorokin, P.B.; Jin, C.; Ni, J.; Kvashnin, A.G.; Kvashnin, D.G.; Lou, J.; Yakobson, B.I. Large scale growth and characterization of atomic hexagonal boron nitride layers. *Nano Lett.* **2010**, *10*, 3209–3215. [CrossRef] [PubMed]

12. Yu, J.; Qin, L.; Hao, Y.; Kuang, S.; Bai, X.; Chong, Y.-M.; Zhang, W.; Wang, E. Vertically aligned boron nitride nanosheets: Chemical vapor synthesis, ultraviolet light emission, and superhydrophobicity. *ACS Nano* **2010**, *4*, 414–422. [CrossRef] [PubMed]

13. Xu, M.; Fujita, D.; Chen, H.; Hanagata, N. Formation of monolayer and few-layer hexagonal boron nitride nanosheets via surface segregation. *Nanoscale* **2011**, *3*, 2854–2858. [CrossRef] [PubMed]

14. Pacile, D.; Meyer, J.; Girit, Ç.; Zettl, A. The two-dimensional phase of boron nitride: Few-atomic-layer sheets and suspended membranes. *Appl. Phys. Lett.* **2008**, *92*, 133107. [CrossRef]

15. Li, L.H.; Chen, Y.; Behan, G.; Zhang, H.; Petravic, M.; Glushenkov, A.M. Large-scale mechanical peeling of boron nitride nanosheets by low-energy ball milling. *J. Mater. Chem.* **2011**, *21*, 11862–11866. [CrossRef]

16. Li, L.H.; Glushenkov, A.M.; Hait, S.K.; Hodgson, P.; Chen, Y. High-efficient production of boron nitride nanosheets via an optimized ball milling process for lubrication in oil. *Sci. Rep.* **2014**, *4*, 7288.

17. Lin, Y.; Williams, T.V.; Connell, J.W. Soluble, exfoliated hexagonal boron nitride nanosheets. *J. Phys. Chem. Lett.* **2009**, *1*, 277–283. [CrossRef]

18. Zhi, C.; Bando, Y.; Tang, C.; Kuwahara, H.; Golberg, D. Large-scale fabrication of boron nitride nanosheets and their utilization in polymeric composites with improved thermal and mechanical properties. *Adv. Mater.* **2009**, *21*, 2889–2893. [CrossRef]

19. Warner, J.H.; Rummeli, M.H.; Bachmatiuk, A.; Büchner, B. Atomic resolution imaging and topography of boron nitride sheets produced by chemical exfoliation. *ACS Nano* **2010**, *4*, 1299–1304. [CrossRef] [PubMed]

20. Zhou, K.G.; Mao, N.N.; Wang, H.X.; Peng, Y.; Zhang, H.L. A mixed-solvent strategy for efficient exfoliation of inorganic graphene analogues. *Angew. Chem. Int. Edit.* **2011**, *50*, 10839–10842. [CrossRef] [PubMed]

21. Coleman, J.N.; Lotya, M.; O'Neill, A.; Bergin, S.D.; King, P.J.; Khan, U.; Young, K.; Gaucher, A.; De, S.; Smith, R.J. Two-dimensional nanosheets produced by liquid exfoliation of layered materials. *Science* **2011**, *331*, 568–571. [CrossRef] [PubMed]

22. Ye, H.; Lu, T.; Xu, C.; Han, B.; Meng, N.; Xu, L. Liquid-phase exfoliation of hexagonal boron nitride into boron nitride nanosheets in common organic solvents with hyperbranched polyethylene as stabilizer. *Macromol. Chem. Phys.* **2018**, *219*, 1700482. [CrossRef]

23. Griffin, A.; Harvey, A.; Cunningham, B.; Scullion, D.; Tian, T.; Shih, C.-J.; Gruening, M.; Donegan, J.F.; Santos, E.J.; Backes, C. Spectroscopic size and thickness metrics for liquid-exfoliated h-BN. *Chem. Mater.* **2018**, *30*, 1998–2005. [CrossRef]

24. Pakdel, A.; Bando, Y.; Golberg, D. Nano boron nitride flatland. *Chem. Soc. Rev.* **2014**, *43*, 934–959. [CrossRef] [PubMed]

25. Ge, Y.; Wang, J.; Shi, Z.; Yin, J. Gelatin-assisted fabrication of water-dispersible graphene and its inorganic analogues. *J. Mater. Chem.* **2012**, *22*, 17619–17624. [CrossRef]

26. Cao, L.; Emami, S.; Lafdi, K. Large-scale exfoliation of hexagonal boron nitride nanosheets in liquid phase. *Mater. Express* **2014**, *4*, 165–171. [CrossRef]

27. Scherrer, P. *Bestimmung der Inneren Struktur und der Größe von Kolloidteilchen Mittels Röntgenstrahlen*; Kolloidchemie Ein Lehrbuch; Springer: Berlin/Heidelberg, Germany, 1912; Chemische Technologie in Einzeldarstellungen.

28. Kim, G.; Jang, A.-R.; Jeong, H.Y.; Lee, Z.; Kang, D.J.; Shin, H.S. Growth of high-crystalline, single-layer hexagonal boron nitride on recyclable platinum foil. *Nano Lett.* **2013**, *13*, 1834–1839. [CrossRef] [PubMed]

29. Gorbachev, R.V.; Riaz, I.; Nair, R.R.; Jalil, R.; Britnell, L.; Belle, B.D.; Hill, E.W.; Novoselov, K.S.; Watanabe, K.; Taniguchi, T. Hunting for monolayer boron nitride: Optical and raman signatures. *Small* **2011**, *7*, 465–468. [CrossRef] [PubMed]

30. Li, L.H.; Cervenka, J.; Watanabe, K.; Taniguchi, T.; Chen, Y. Strong oxidation resistance of atomically thin boron nitride nanosheets. *ACS Nano* **2014**, *8*, 1457–1462. [CrossRef] [PubMed]

31. Zhu, W.; Gao, X.; Li, Q.; Li, H.; Chao, Y.; Li, M.; Mahurin, S.M.; Li, H.; Zhu, H.; Dai, S. Controlled gas exfoliation of boron nitride into few-layered nanosheets. *Angew. Chem. Int. Edit.* **2016**, *55*, 10766–10770. [CrossRef] [PubMed]

32. Cai, Q.; Scullion, D.; Falin, A.; Watanabe, K.; Taniguchi, T.; Chen, Y.; Santos, E.J.; Li, L.H. Raman signature and phonon dispersion of atomically thin boron nitride. *Nanoscale* **2017**, *9*, 3059–3067. [CrossRef] [PubMed]

33. Sainsbury, T.; Satti, A.; May, P.; Wang, Z.; McGovern, I.; Gun'ko, Y.K.; Coleman, J. Oxygen radical functionalization of boron nitride nanosheets. *J. Am. Chem. Soc.* **2012**, *134*, 18758–18771. [CrossRef] [PubMed]

nanomaterials

MDPI

Article

Effective Mechanical Properties and Thickness Determination of Boron Nitride Nanosheets Using Molecular Dynamics Simulation

Venkatesh Vijayaraghavan and Liangchi Zhang *

Laboratory for Precision and Nano Processing Technologies, School of Mechanical and Manufacturing Engineering, The University of New South Wales, Sydney, NSW 2052, Australia; venkatesh.vijayaraghavan@unsw.edu.au
* Correspondence: liangchi.zhang@unsw.edu.au; Tel.: +61-2-9385-6078

Received: 3 July 2018; Accepted: 17 July 2018; Published: 19 July 2018

Abstract: Research in boron nitride nanosheets (BNNS) has evoked significant interest in the field of nano-electronics, nanoelectromechanical (NEMS) devices, and nanocomposites due to its excellent physical and chemical properties. Despite this, there has been no reliable data on the effective mechanical properties of BNNS, with the literature reporting a wide scatter of strength data for the same material. To address this challenge, this article presents a comprehensive analysis on the effect of vital factors which can result in variations of the effective mechanical properties of BNNS. Additionally, the article also presents the computation of the correct wall thickness of BNNS from elastic theory equations, which is an important descriptor for any research to determine the mechanical properties of BNNS. It was predicted that the correct thickness of BNNS should be 0.106 nm and the effective Young's modulus to be 2.75 TPa. It is anticipated that the findings from this study could provide valuable insights on the true mechanical properties of BNNS that could assist in the design and development of efficient BN-based NEMS devices, nanosensors, and nanocomposites.

Keywords: boron nitride nanosheet; molecular dynamics; thickness; mechanical strength; vacancy defects

---

1. Introduction

Research on boron nitride nanosheets (BNNS) has evoked great prominence in recent days owing to its unique physical and electronic properties [1,2]. BNNS is the best example of a 2D single-layer compound nanomaterial, consisting of boron and nitrogen atoms in equal numbers in a hexagonal lattice arrangement. While the mechanical strength of BNNS is not as high compared to that of its elemental counterpart—graphene [3], it possesses better thermal and oxidation resistance [4]. This makes BNNS an attractive alternative to graphene for applications under extreme conditions, such as nanocomposites [5,6], nanoelectronics [7,8], and nanoelectromechanical (NEMS) devices [9,10]. This creates a major initiative in investigating the mechanical strength of BNNS, which will provide valuable information for the design of next-generation BNNS-based nano-devices and components.

Previous studies investigating the mechanics of BNNS by experiments mainly focused on reporting the Young's modulus of BNNS. The reported Young's modulus of BNNS was scattered and was found to be dependent on the method of fabrication of the BNNS. For instance, Falin et al. [1] and Bosak et al. [11] and devised the fabrication of BNNS by exfoliation from BN crystals and found the Young's modulus of BNNS to be 0.865 and 0.811 TPa, respectively. However, Song et al. [12] predicted that the BNNS fabricated by the chemical vapour deposition (CVD) process, yielded a Young's modulus of only about 0.334 TPa. The low strength was attributed to the inherent defects and grain boundaries resulting from the CVD process. Kim et al. [13], on the other hand, adopted the CVD process using an iron foil with a borazine precursor to synthesize high-quality BNNS. They reported that the Young's modulus of BNNS can be substantially higher for the case of BN with little or no inherent defects and obtained a Young's modulus of 1.16 ± 0.1 TPa. A similar observation was also reported for the case of boron nitride nanotubes (BNNTs) by Chopra and Zettl [14] who devised a water cooled arc for synthesizing pure BNNTs. They found that the Young's modulus of BNNS is 1.22 ± 0.24 TPa. Suryavanshi et al. [15] measured the Young's modulus of 18 different BNNTs with varying lengths and diameters and also reported a wide scatter of the Young's modulus, varying from 0.505–1.031 TPa. These studies also adopted varying measurement techniques, such as atomic force microscopy (AFM), transmission electron microscopy (TEM), inelastic X-ray scattering (IXS) technique, etc., to name a few. The measurement errors in these techniques might have also contributed to the diverse range of strength data of BNNS available in the literature. A summary of the mechanical strength of BNNS as reported from the abovementioned experimental studies are presented in Table 1.

Table 1. Mechanical properties of BNNS and BNNTs determined by experiments.

Experimental Method	Young's Modulus (TPa)
Nanoindentation measurement of few layer BNNS exfoliated from single crystal BN [1].	0.865 ± 0.073
IXS of BNNS crystal synthesized from Ba-B-N catalyst system under high temperature and pressure [11].	0.811
Nanoindentation measurement on defective BNNS synthesized by CVD from bulk BN crystal [12].	0.334 ± 0.024
AFM measurement on high quality BNNS synthesized from borazine precursor using CVD process [13].	1.16 ± 0.1
Thermal assisted vibration of cantilevered BNNT observed using TEM [14].	1.22 ± 0.24
Electric-field-induced technique to apply sinusoidal signal which induces vibration in BNNT [15].	0.505–1.031

It is evident from the above experimental studies that the variation in the Young's modulus and the mechanical strength of BNNS can be influenced by many factors. Computational modelling has emerged as an effective means of studying the influence of various parameters, such as defects, geometry, and lattice orientation on the strength data of BNNS. Molecular dynamics (MD) studies [16–18] showed that the Young's modulus of BNNS is highly sensitive to defects in BNNS lattice. Similar conclusions were also obtained from density functional theory (DFT) analysis by Wang et al. [19]. Another advantage of deploying computational model is that the lattice parameters or geometry of BNNS can be easily modified and the resulting mechanical strength can be estimated. For instance, Le [20] adopted a molecular mechanics (MM) model for BNNS undergoing tensile loading, and found that BNNS loaded in armchair direction exhibits lower tensile strength. This observation was also confirmed by MD simulation results of Mortazavi and Rémond [21] and DFT analysis by Wu et al. [22]. In addition to defects and lattice orientation, the strength of BNNS was also reported to be strongly influenced by temperature. Adopting a quasi-harmonic approximation (QHA) model, Mirnezhad et al. [23] showed that the Young's modulus of BNNS is highly sensitive to temperature, reaching a stable value at elevated temperature. Other computational approaches, such as hybrid Tersoff-Brenner (T-B) [24] and continuum-lattice (C-L) [25] models, also reported similar results. Some computational studies also analysed the strength variation of boron nitride nanotubes (BNNTs) [26] and their strength comparison with carbon nanotubes or graphene [27,28]. Most of these studies reported the Young's modulus and mechanical strength of BNNS by considering the wall thickness of BNNS to be around 3.3 to 3.4 Å. Some studies also reported thickness-independent mechanical strength descriptors, such as the axial stiffness and bending stiffness from conventional modelling techniques, such as MD simulations [29,30] or other techniques, such as atomistic-finite element modelling (FEM) [31] or the discrete media homogenization (DMH) technique [32]. A consolidation of all mechanical properties of BNNS investigated by various computational approaches is presented in Table 2.

Table 2. Mechanical properties of BNNS and BNNTs by computational modelling.

Technique	Temperature (K)	Young's Modulus (TPa)	Axial Stiffness (TPa nm)	Bending Stiffness (eV)
Tersoff potential [16]	300	0.930	NA	NA
Tersoff potential [17]	NA	0.730–0.890	0.248–0.292	NA
Tersoff potential [18]	0–2000	0.398–0.720	NA	NA
DFT calculation [19]	NA	NA	0.293–0.311	NA
Mechanics model [20]	0	NA	0.332	NA
Tersoff potential [21]	300	0.800–0.850	0.264–0.280	NA
DFT calculation [22]	NA	0.760–1.055	NA	0.95
DFT-QHA model [23]	0–1000	NA	0.278–0.283	NA
T-B potential [24]	300	0.881	NA	NA
Continuum model [25]	NA	0.900–1.000	NA	NA
Tight binding [26]	NA	NA	0.284–0.310	NA
MM-DFT model [27]	NA	0.83	0.282	1.74
DFT calculation [28]	NA	0.700–0.830	NA	NA
Tersoff potential [29]	0	NA	0.267	NA
Tersoff-like model [30]	300	NA	NA	1.5–1.7
Atomistic-FEM [31]	NA	NA	0.240–0.315	NA
DMH technique [32]	NA	NA	0.267	NA
Tersoff potential [33]	NA	0.295–0.695	NA	0.22–0.56
MM model [34]	NA	NA	0.260–0.269	NA
Ab initio [35]	NA	NA	0.271	1.29
DFT calculation [36]	NA	NA	0.279	NA
Modified T-B [37]	NA	0.982–1.113	NA	NA
Tersoff potential [38]	300	0.716	NA	NA
Tersoff potential [39]	0	0.749–0.770	0.248–0.258	NA

From the literature studies presented above, it is possible to map the effect of individual factors on the mechanical strength of BNNS. Table 3 presents the effect of various factors which can influence the mechanical characteristics of BNNS. This offers a quick glance of the dominant factors which results in the variation of the reported strength data of BNNS. It is also possible to determine the knowledge gaps on the existing studies on the mechanics of BNNS. For instance, while it is evident that the strength of BNNS is adversely affected by the presence of defects and increasing temperature, superior mechanical properties can be obtained by orienting the BNNS along a zigzag direction. While the effect of individual factors, such as defects and temperature, has been well documented, none of the abovementioned studies focused on analysing the effect of BNNS geometry, the position of defects, and the combined effect of two or more factors in influencing the strength data of BNNS. Additionally, almost all of the existing studies in the literature reported the mechanical strength and Young's modulus of BNNS by assuming the thickness of BNNS to be 3.3 to 3.4 Å, which is the inter-layer separation distance of graphene. This assumption is not true given that the BNNS exhibits a discrete hexagonal lattice arrangement of atoms and, hence, its wall thickness is not well defined. Estimating the correct wall thickness is very crucial to determining the effective Young's modulus and mechanical properties of BNNS at the nanoscale. Assuming an incorrect wall thickness leads to a scatter of the Young's modulus and mechanical strength data of BNNS, this leads to a critical knowledge gap on exploiting these nanomaterials for high-strength applications.

Table 3. Relationship mapping of the effect of various factors on the mechanical strength of BNNS.

Factors	Tensile Strength of BNNS
Temperature	Decreases
Defect concentration	Decreases
Geometry	Unknown
Defect position	Unknown
Defects and Temperature	Unknown
Loading direction	Superior in zigzag direction

Motivated by the above research questions, this article aims to provide a comprehensive analysis on the mechanics of BNNS under tensile loading conditions. To this end, the critical factors which can result in a variation of strength data of BNNS are identified and the extent to which the strength data is varied is determined first. Additionally, this article also addresses the need to compute the correct wall thickness of BNNS, and thereby its effective Young's modulus and mechanical strength at the nanoscale. The correct wall thickness of the BNNS is estimated by adopting the Vodenitcharova-Zhang [40] and Wang-Zhang [41] criteria. Once the correct wall thickness is determined, the effective Young's modulus and the mechanical strength of BNNS is then calculated and presented.

2. Computational Model

This article focuses on the mechanics of a single-layer BNNS using MD simulation. The simulations are performed on the large scale atomic/molecular massively parallel simulator (LAMMPS) package of March 2017 version, developed by Sandia National Laboratories, Livermore, CA, USA [42]. The modified Tersoff potential [43,44] with optimized parameters defined by Kinaci et al. [45] is used to describe the interactions between the boron and nitrogen atoms of BNNS. The Tersoff potential, with its precise parameters, has the ability to accurately match the experimental results with density functional theory calculations, while also ensuring the computational efficiency for large-scale atomic systems [46]. In addition, the Tersoff potential has also been successfully used in previous studies on computational modelling of BNNS [21,47–49]. The complete details of this potential function with associated parameters can be found in [18].

The studies described in this paper analyses the effect of geometry, loading direction, defects, and temperature on the mechanical strength of BNNS. The effect of geometry is considered by suitably modifying the aspect ratio (ratio of length to the width) of the BNNS in zigzag and armchair directions. The effect of concentration and position of vacancy defects is studied by constructing various concentrations axial or transverse defects along the direction of loading of the BNNS. The temperature factor is investigated by subjecting the BNNS structure to tensile loading at 300, 600, and 900 K. At the beginning, the BNNS is equilibrated at the specific temperature, after which the boundary atoms of the BNNS are fixed and subjected to constant outward displacement to simulate tension (Figure 1). The BNNS is again equilibrated at every 1000 time steps to relax the structure, after which the readings are recorded and the procedure is repeated until the BNNS fails under tension.

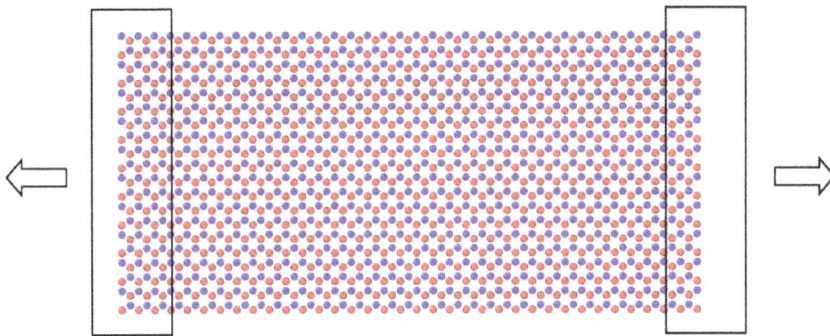

Figure 1. Simulation of BNNS under tensile loading. The atoms enclosed inside the black rectangle at either ends of BNNS is subjected to tensile loading. The loading direction is indicated by the arrows. Atoms depicted in ochre are boron and atoms depicted in blue are nitrogen.

3. Results and Discussion

3.1. Validation of the Simulation Model

The simulation model adopted in the present study is validated by considering an approximate square shaped BNNS of dimension 62.38 Å × 60.27 Å consisting of 700 boron and nitrogen atoms. The BNNS is loaded in tension along the zigzag direction as depicted in Figure 1 at 300 K. The plot of the force and strain energy per atom measured against the tensile strain is illustrated in Figure 2. The plot shows that the tensile force varies almost linearly with strain, ε, until $\varepsilon = 0.05$. After this, the force shows a parabolic variation until it reaches a maximum value of 198.14 nN before undergoing failure as indicated by the spontaneous drop of strain energy. The observed tensile strain of $\varepsilon = 0.27$ from the present MD simulation is in good agreement with the finite element prediction value of 0.257 by Le and Nguyen [39] who adopted a square-shaped model of BNNS. The computed axial stiffness of the BNNS (defined as twice the coefficient of the second degree term of the strain energy polynomial curve) is 277.4 J/m^2 which is comparable with the ab initio prediction of 271 J/m^2 [35] and MD prediction of 267 J/m^2 [29]. Hence, the above confirmation study validates the accuracy of the simulation model adopted in the study.

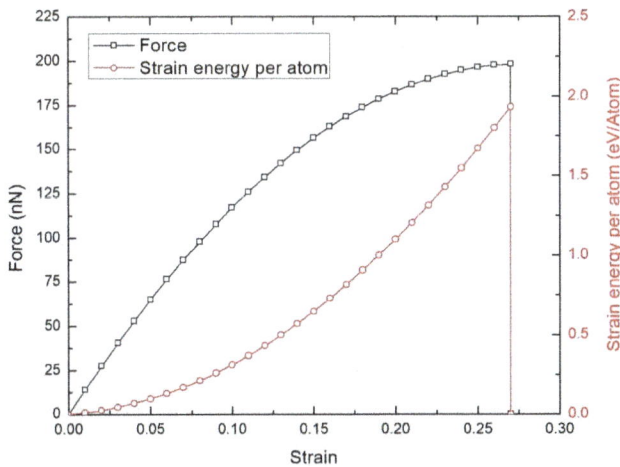

Figure 2. Force and strain energy graph of BNNS under tensile loading at 300 K.

3.2. Effect of Geometry and Tensile Loading Direction

The effect of geometry and tensile loading direction on the mechanical characteristics of BNNS is described in this section. It is useful to note that previous studies have modelled BNNS with varying geometry, such as rectangular [18,21], square shaped [39], or even as a circle [50], and have reported varying data of tensile strength of single-layer BNNS. The strength data of BNNS should be ideally investigated by standardizing the variation in geometry. This is accomplished in this work by varying the aspect ratio (ratio of length, L to the width, W) of BNNS, while maintaining almost the same number of boron and nitrogen atoms. The effect of the loading direction is considered by tensile loading of BNNS along either armchair or zigzag directions. The geometry and atomic configuration of armchair and zigzag BNNS considered in the study is presented in Tables 4 and 5, respectively. The plot of the maximum tensile force for armchair and zigzag BNNS as a function of the aspect ratio at 300 K is shown in Figure 3. The plot shows that the BNNS exhibits superior tensile strength when loaded in the zigzag direction, which is consistent with the previous literatures. Additionally, the maximum tensile force of BNNS decreases with the increasing aspect ratio. This is because a small aspect ratio BNNS is wider than a larger aspect ratio BNNS. In the present study, as there is minor variation in the atom numbers of BNNS across aspect ratios, a wider BNNS has more atomic bonds which can resist the tensile loading. Furthermore, the loading characteristics tend to stabilize with the increasing aspect ratio of the BNNS. Hence, the mechanical strength of the BNNS can be effectively modified by varying the system geometry.

Table 4. The atomic configuration of single layer BNNS loaded in the zigzag direction.

Aspect Ratio (L/W)	BNNS Dimensions (L × W)	Total Number of Atoms
1.0	62.38 Å × 60.27 Å	1408
2.0	89.11 Å × 42.63 Å	1420
3.0	104.39 Å × 33.81 Å	1328
4.0	120.94 Å × 29.40 Å	1344

Table 5. The atomic configuration of single layer BNNS loaded in the armchair direction.

Aspect Ratio (L/W)	BNNS Dimensions (L × W)	Total Number of Atoms
1.0	62.48 Å × 61.11 Å	1450
2.0	86.73 Å × 43.28 Å	1400
3.0	104.37 Å × 33.10 Å	1344
4.0	119.81 Å × 30.55 Å	1374

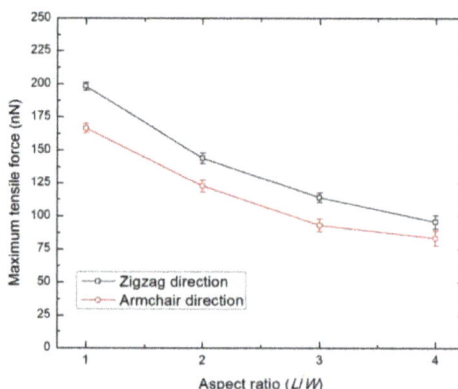

Figure 3. Tensile loading characteristics of single layer BNNS of varying aspect ratios at 300 K.

The snapshots of the single layer BNNS of aspect ratio 2.0 undergoing tensile loading at 300 K is depicted in Figure 4. The initial equilibration at 300 K results in some wrinkles on the BNNS structure. The application of tensile displacement results in elongation and increase in strain energy of the atoms, which results in tensile failure of the BNNS sheet marked by fracture and segmentation.

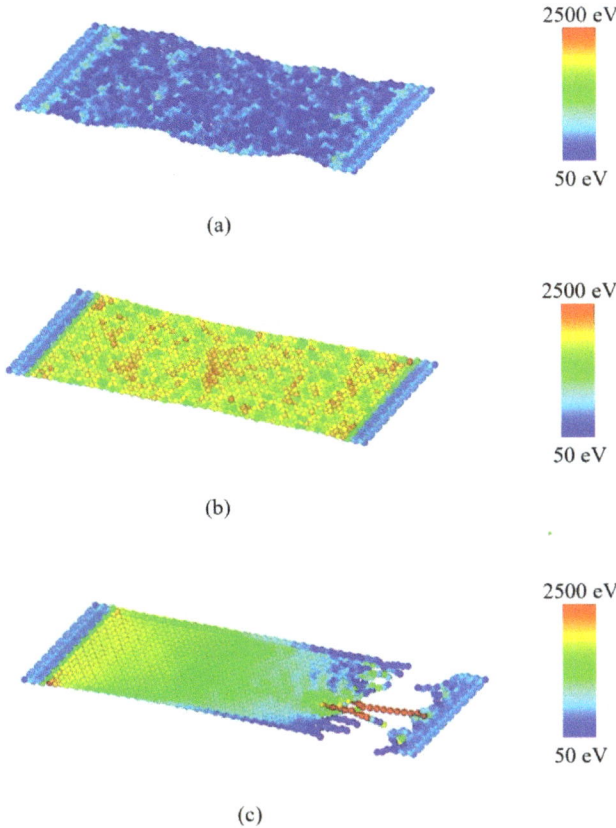

(a)

(b)

(c)

Figure 4. Tensile loading stages of single-layer BNNS at 300 K at (**a**) $\varepsilon = 0.0$; (**b**) $\varepsilon = 0.12$; and (**c**) $\varepsilon = 0.25$.

3.3. Effect of the Concentration and Position of Vacancy Defect

The influence of vacancy defects on the tensile loading characteristics of BNNS is investigated by considering a BNNS sheet of aspect ratio 2.0, loaded in armchair and zigzag directions at 300 K. The vacancy defect is constructing by removing a set of covalently bonded boron and nitrogen atoms from the BNNS lattice. Figure 5 shows that increasing the concentration of defects deteriorates the mechanical strength of BNNS. It is, however, intriguing to note that the extent of the influence exerted by the defects has strong dependency on their position or placement in the BNNS. For instance, the maximum tensile force for an armchair BNNS drops by over 43% when the vacancy defect concentration along the transverse direction (i.e., perpendicular to direction of loading) is increased to 6. However, the corresponding reduction in the maximum tensile force when the defects are located in the axial direction is only about 20%. This is an important characteristic which should be taken note of while synthesizing BNNS for composite loading or high-strength applications. Hence, the vacancy defects along the transverse direction must be minimized as much as possible to avoid the rapid deterioration in the mechanical properties of BNNS.

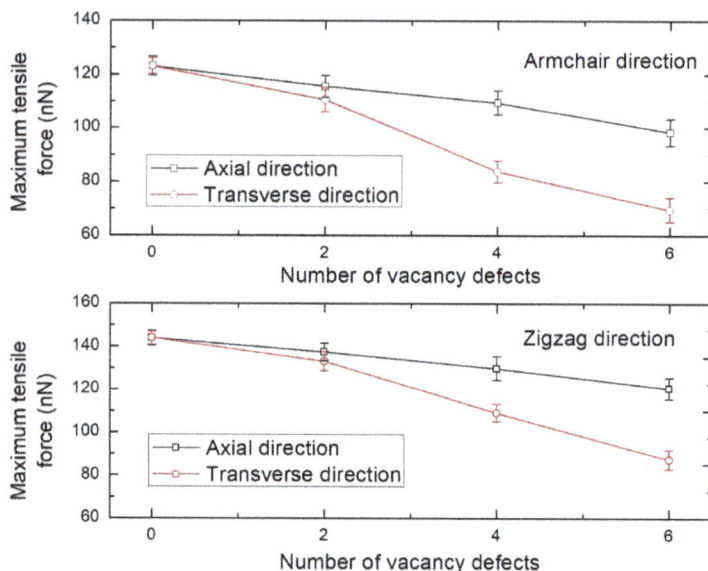

Figure 5. Tensile loading characteristics of single layer BNNS with vacancy defects along the axial and transverse directions at 300 K.

3.4. Effect of Temperature and Vacancy Defects

The previous studies presented in the introduction indicated that the strength of BNNS is strongly affected by temperature and the existence of defects. However, the strength of defective BNNS has not been tested in elevated temperatures, or vice versa, which will provide a comprehensive understanding of the interaction of these two dominant factors. Figure 6 depicts the variation in tensile loading characteristics of BNNS with increasing defect concentration at various temperatures. It can be witnessed that the increase in temperature decreases the maximum tensile force of the BNNS, which is attributed to the increase in the thermal stress on B–N bonds. However, it is also interesting to note that introducing the vacancy defects in the BNNS structure seems to mitigate the weakening effect of BNNS caused due to the rise in temperature (Table 6). For instance, the reduction in the maximum tensile force of a pristine BNNS when the temperature is increased from 300 to 900 K is 14.25%, while the corresponding drop for a BNNS with six vacancy defects is found to be lowered to 10.16%. The loss of B–N bonds formed in BNNS due to the presence of vacancy defects enhances the mobility of the atoms at higher temperature which results in the lowering of associated thermal stress in the BNNS structure. This observation is also consistent with the mechanics of the graphene sheet [51] and buckling of CNTs [52] analysed at higher temperatures. Hence, BNNS with a higher concentration of vacancy defects resists the drop in tensile loading characteristics due to temperature increases. This could be an important factor which could be exploited while synthesizing BNNS for high strength applications under elevated temperatures. Another useful feature which can be deduced from this study is that the drop in loading characteristics of BNNS due to temperature variation is not as pronounced when compared to that of graphene as reported by authors' previous study [51]. Hence, this investigation supports the fact that BNNS exhibits better thermal stability and can be used for fabricating temperature-resistant nanoscale devices and nanocomposites.

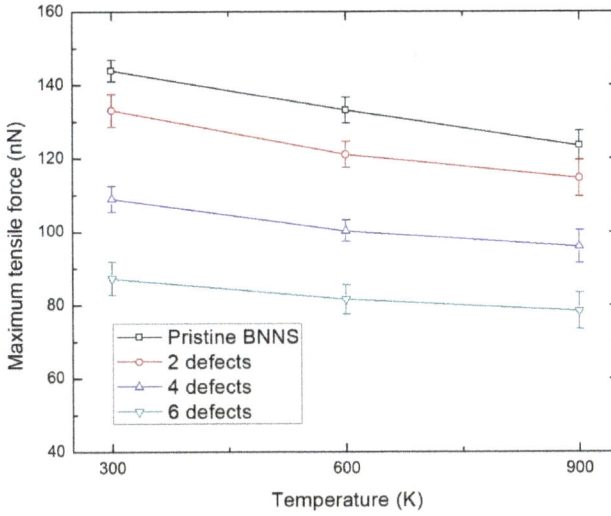

Figure 6. Tensile loading characteristics of BNNS at various temperatures with vacancy defects.

Table 6. Percentage reduction of the maximum tensile force of BNNS with defects when temperature is increased from 300 to 900 K.

Number of Defects	Reduction of Maximum Tensile Force (%)
0	14.25
2	13.83
4	11.97
6	10.16

4. Determination of Thickness and the Young's Modulus of BNNS

Almost all of the previous studies on computational modelling of BNNS have assumed the thickness to be 3.4 Å—the inter-layer separation between two graphene sheets. This yields the Young's modulus to be about 0.6–0.9 TPa. The application of the same thickness to estimate the mechanical characteristics of BNNS is questionable since the effective thickness of graphene itself was computed to be between 0.06 to 0.1 nm [41,53].

To overcome this hurdle, the effective thickness of BNNS is determined in this work based on the well-established Vodenitcharova-Zhang [40] and Wang-Zhang [41] criteria. In so doing, the axial stiffness, K, and bending stiffness, D, are firstly determined without using E and h values. In the atomistic simulation of BNNS, the axial stiffness K is defined as [41]:

$$\begin{cases} K = \frac{1}{A} \cdot \frac{\partial^2 W_a}{\partial \varepsilon^2} \\ W = a_0 + a_1 \varepsilon + a_2 \varepsilon^2 + a_3 \varepsilon^3 + \cdots \end{cases} \tag{1}$$

where W_a is the strain energy of the BNNS structure under axial loading, A is the surface area of the BNNS, a_j ($j = 0,1,2,3, \ldots$) is the coefficient of the fitted polynomial of W_a in terms of strain, and ε derived from the strain energy-strain plot.

The bending stiffness, D, of BNNS is determined by the energy required in rolling up the BNNS surface to form a BNNT (see Figure 7). D is defined mathematically as [41]:

$$\begin{cases} D = \frac{1}{A} \cdot \frac{\partial^2 W_b}{\partial \kappa^2} \\ W_b = b_0 + b_1 \kappa + b_2 \kappa^2 + b_3 \kappa^3 + \cdots \end{cases} \tag{2}$$

where W_b is the energy of the BNNS structure during bending process to form a BNNT, b_j ($j = 0,1,2,3,$ ...) is the coefficient of the fitted polynomial of W_b in terms of curvature, and κ is derived from the energy-curvature plot.

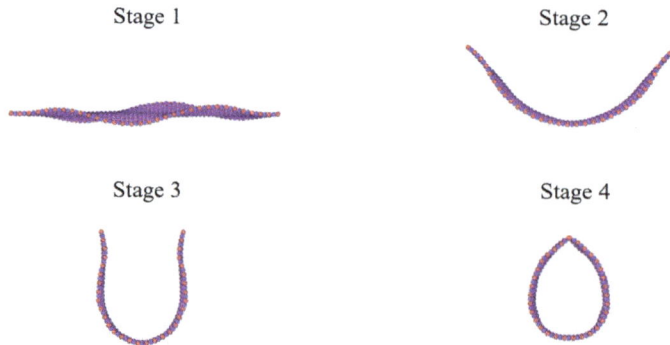

Stage 1

Stage 2

Stage 3

Stage 4

Figure 7. Stages of bending in BNNS to form a BNNT.

Based on elastic theory, the K and D values are defined in terms of E and h as [41]:

$$K = \frac{Eh}{1 - \upsilon^2} \approx Eh \tag{3}$$

$$D = \frac{Eh^3}{12(1 - \upsilon^2)} \approx \frac{Eh^3}{12} \tag{4}$$

The above equations are then solved to determine the unique values of E and h, which is sufficient to satisfy the axial stiffness and bending stiffness of BNNS. In addition, it is also necessary for the resultant thickness to be smaller than that of the atomic diameter, since the cross-section of the sheet only consists of discrete atoms connected by bonds, as opposed to a continuous wall of atoms.

The variation of thickness with the Young's modulus is plotted on a single *E-h* coordinate plane using Equations (3) and (4), as shown in Figure 8. In the present study, the average K and D values of BNNS at various aspect ratios were obtained as 285.7 J/m^2 and 1.785 eV, respectively. These values are in good agreement with the K and D values computed from various numerical approaches, as illustrated in Table 1. From Figure 7, the correct thickness of BNNS is determined by the intersection of the K and D curves, while also satisfying the Vodenitcharova-Zhang necessary criterion [40]. Hence, the correct effective thickness of BNNS is $h \approx 0.106$ nm and the Young's modulus ≈ 2.75 TPa.

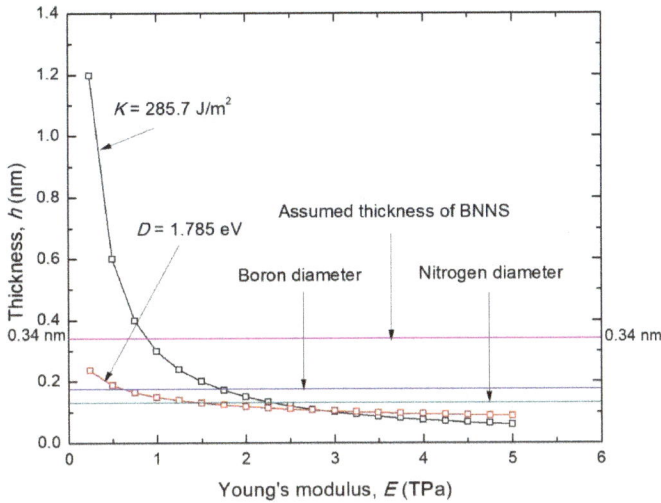

Figure 8. Determination of the correct thickness and Young's modulus of BNNS from the intersection of axial stiffness and bending stiffness curves on the *E-h* coordinate plane.

The resulting Young's modulus is higher than that of the previously reported estimates using computer simulation studies. Hence, the following checks can be conducted to confirm the validity of the computed Young's modulus:

(1) The correct wall thickness for graphene was estimated to be about 0.10 nm [41,54]. Since BNNS is morphologically similar to the graphene sheet, the BNNS thickness of 0.106 nm is closely comparable to the thickness of the graphene sheet.

(2) For a thickness of 3.4 Å, the Young's modulus of the BNNS reported by computational studies is lower than that of graphene, and should be valid regardless of any thickness considered. As the computed modulus of BNNS (2.75 TPa) is lower than the correct modulus of graphene, which is reported to be 3.4–3.5 TPa [41,55], the above findings can be validated.

Using the effective thickness of 0.106 nm and the maximum tensile force values reported for the armchair and zigzag BNNS in Section 3.2, the mechanical strength of the BNNS with varying aspect ratios are computed and presented in Table 7.

Table 7. Effective mechanical strength of armchair and zigzag BNNS at various aspect ratios.

Armchair BNNS		Zigzag BNNS	
Aspect Ratio	**Mechanical Strength (GPa)**	**Aspect Ratio**	**Mechanical Strength (GPa)**
1.0	254.31	1.0	302.75
2.0	266.82	2.0	318.51
3.0	259.33	3.0	317.70
4.0	266.21	4.0	306.67

5. Conclusions

Mechanical loading characteristics of BNNS under tensile loading conditions have been comprehensively analysed in this work. A detailed literature review has been conducted to consolidate the effect of various factors which can influence the quantification of the mechanical properties of BNNS. Based on the literature consolidation, it is identified that the influence of geometry, defect position, and the combination of defects and temperature on the property characterization of BNNS must be investigated. The variation in the system geometry was standardized across BNNS by maintaining almost a similar number of atoms in the sheet. Through this, it was found that a smaller aspect ratio of BNNS exhibits better tensile loading characteristics. Furthermore, while increasing the defect concentration, itself, can deteriorate the mechanical strength of BNNS, the extent of the reduction was found to have a strong dependency on the position or placement of the defects. The study also revealed interesting phenomena that these vacancy defects can control the decline in tensile loading characteristics of BNNS due to elevated temperatures. Hence, it would be favourable to include vacancy defects in BNNS for high-temperature applications, albeit placing the defects along the direction of loading of BNNS. Finally, the effective Young's modulus of the BNNS is also estimated by computing the correct wall thickness based on elastic theory equations. It is anticipated that the comprehensive analysis presented in this work will provide valuable information for the fabrication of BNNS-based NEMS, nanoscale devices, and nanocomposites.

Author Contributions: Conceptualization: V.V.; formal analysis: V.V.; investigation: V.V.; methodology: V.V.; supervision: L.Z.; writing—original draft: V.V.; writing—review and editing: L.Z.

Funding: This research received no external funding.

Acknowledgments: The first author, V. Vijayaraghavan acknowledges the faculty-supported fellowship by the University of New South Wales, Australia.

Conflicts of Interest: The authors declare no conflict of interest.

References

1. Falin, A.; Cai, Q.; Santos, E.J.G.; Scullion, D.; Qian, D.; Zhang, R.; Yang, Z.; Huang, S.; Watanabe, K.; Taniguchi, T.; et al. Mechanical properties of atomically thin boron nitride and the role of interlayer interactions. *Nat. Commun.* **2017**, *8*, 15818. [CrossRef] [PubMed]
2. Zeng, H.; Zhi, C.; Zhang, Z.; Wei, X.; Wang, X.; Guo, W.; Bando, Y.; Golberg, D. "White graphenes": Boron nitride nanoribbons via boron nitride nanotube unwrapping. *Nano Lett.* **2010**, *10*, 5049–5055. [CrossRef] [PubMed]
3. Zhang, J.; Wang, C. Mechanical properties of hybrid boron nitride-carbon nanotubes. *J. Phys. D Appl. Phys.* **2016**, *49*, 15. [CrossRef]
4. Golberg, D.; Bando, Y.; Huang, Y.; Terao, T.; Mitome, M.; Tang, C.; Zhi, C. Boron nitride nanotubes and nanosheets. *ACS Nano* **2010**, *4*, 2979–2993. [CrossRef] [PubMed]
5. Isarn, I.; Ramis, X.; Ferrando, F.; Serra, A. Thermoconductive thermosetting composites based on boron nitride fillers and thiol-epoxy matrices. *Polymers* **2018**, *10*, 277. [CrossRef]
6. Yu, J.; Zhao, W.; Wu, Y.; Wang, D.; Feng, R. Tribological properties of epoxy composite coatings reinforced with functionalized C-BN and H-BN nanofillers. *Appl. Surf. Sci.* **2018**, *434*, 1311–1320. [CrossRef]
7. Shahrokhi, M.; Mortazavi, B.; Berdiyorov, G.R. New two-dimensional boron nitride allotropes with attractive electronic and optical properties. *Solid State Commun.* **2017**, *253*, 51–56. [CrossRef]
8. Tao, X.; Zhang, L.; Zheng, X.; Hao, H.; Wang, X.; Song, L.; Zeng, Z.; Guo, H. *h*-BN/graphene van der Waals vertical heterostructure: A fully spin-polarized photocurrent generator. *Nanoscale* **2018**, *10*, 174–183. [CrossRef] [PubMed]
9. Arutt, C.N.; Alles, M.L.; Liao, W.; Gong, H.; Davidson, J.L.; Schrimpf, R.D.; Reed, R.A.; Weller, R.A.; Bolotin, K.; Nicholl, R.; et al. The study of radiation effects in emerging micro and nano electro mechanical systems (M and NEMS). *Semicond. Sci. Technol.* **2017**, *32*, 1. [CrossRef]
10. Garel, J.; Zhao, C.; Popovitz-Biro, R.; Golberg, D.; Wang, W.; Joselevich, E. BCN nanotubes as highly sensitive torsional electromechanical transducers. *Nano Lett.* **2014**, *14*, 6132–6137. [CrossRef] [PubMed]
11. Bosak, A.; Serrano, J.; Krisch, M.; Watanabe, K.; Taniguchi, T.; Kanda, H. Elasticity of hexagonal boron nitride: Inelastic X-ray scattering measurements. *Phys. Rev. B Condens. Matter Mater. Phys.* **2006**, *73*, 041402(R). [CrossRef]

12. Song, L.; Ci, L.; Lu, H.; Sorokin, P.B.; Jin, C.; Ni, J.; Kvashnin, A.G.; Kvashnin, D.G.; Lou, J.; Yakobson, B.I.; et al. Large scale growth and characterization of atomic hexagonal boron nitride layers. *Nano Lett.* **2010**, *10*, 3209–3215. [CrossRef] [PubMed]

13. Kim, S.M.; Hsu, A.; Park, M.H.; Chae, S.H.; Yun, S.J.; Lee, J.S.; Cho, D.H.; Fang, W.; Lee, C.; Palacios, T.; et al. Synthesis of large-area multilayer hexagonal boron nitride for high material performance. *Nat. Commun.* **2015**, *6*, 8662. [CrossRef] [PubMed]

14. Chopra, N.G.; Zettl, A. Measurement of the elastic modulus of a multi-wall boron nitride nanotube. *Solid State Commun.* **1998**, *105*, 297–300. [CrossRef]

15. Suryavanshi, A.P.; Yu, M.F.; Wen, J.; Tang, C.; Bando, Y. Elastic modulus and resonance behavior of boron nitride nanotubes. *Appl. Phys. Lett.* **2004**, *84*, 2527–2529. [CrossRef]

16. Eshkalak, K.E.; Sadeghzadeh, S.; Jalaly, M. Mechanical properties of defective hybrid graphene-boron nitride nanosheets: A molecular dynamics study. *Comput. Mater. Sci.* **2018**, *149*, 170–181. [CrossRef]

17. Griebel, M.; Hamaekers, J.; Heber, F. A molecular dynamics study on the impact of defects and functionalization on the young modulus of boron-nitride nanotubes. *Comput. Mater. Sci.* **2009**, *45*, 1097–1103. [CrossRef]

18. Li, N.; Ding, N.; Qu, S.; Liu, L.; Guo, W.; Wu, C.M.L. Mechanical properties and failure behavior of hexagonal boron nitride sheets with nano-cracks. *Comput. Mater. Sci.* **2017**, *140*, 356–366. [CrossRef]

19. Wang, H.; Ding, N.; Zhao, X.; Wu, C.M.L. Defective boron nitride nanotubes: Mechanical properties, electronic structures and failure behaviors. *J. Phys. D Appl. Phys.* **2018**, *51*, 12. [CrossRef]

20. Le, M.Q. Prediction of young's modulus of hexagonal monolayer sheets based on molecular mechanics. *Int. J. Mech. Mater. Des.* **2015**, *11*, 15–24. [CrossRef]

21. Mortazavi, B.; Rémond, Y. Investigation of tensile response and thermal conductivity of boron-nitride nanosheets using molecular dynamics simulations. *Phys. E Low-Dimens. Syst. Nanostruct.* **2012**, *44*, 1846–1852. [CrossRef]

22. Wu, J.; Wang, B.; Wei, Y.; Yang, R.; Dresselhaus, M. Mechanics and mechanically tunable band gap in single-layer hexagonal boron-nitride. *Mater. Res. Lett.* **2013**, *1*, 200–206. [CrossRef]

23. Mirnezhad, M.; Ansari, R.; Shahabodini, A. Temperature effect on young's modulus of boron nitride sheets. *J. Therm. Stresses* **2013**, *36*, 152–159. [CrossRef]

24. Han, T.; Luo, Y.; Wang, C. Effects of temperature and strain rate on the mechanical properties of hexagonal boron nitride nanosheets. *J. Phys. D Appl. Phys.* **2014**, *47*, 2. [CrossRef]

25. Oh, E.S. Elastic properties of boron-nitride nanotubes through the continuum lattice approach. *Mater. Lett.* **2010**, *64*, 859–862. [CrossRef]

26. Hernández, E.; Goze, C.; Bernier, P.; Rubio, A. Elastic properties of single-wall nanotubes. *Appl. Phys. A Mater. Sci. Process.* **1999**, *68*, 287–292. [CrossRef]

27. Ansari, R.; Mirnezhad, M.; Sahmani, S. Prediction of chirality- and size-dependent elastic properties of single-walled boron nitride nanotubes based on an accurate molecular mechanics model. *Superlattices Microstruct.* **2015**, *80*, 196–205. [CrossRef]

28. Akdim, B.; Pachter, R.; Duan, X.; Adams, W.W. Comparative theoretical study of single-wall carbon and boron-nitride nanotubes. *Phys. Rev. B Condens. Matter Mater. Phys.* **2003**, *67*, 245404. [CrossRef]

29. Los, J.H.; Kroes, J.M.H.; Albe, K.; Gordillo, R.M.; Katsnelson, M.I.; Fasolino, A. Extended tersoff potential for boron nitride: Energetics and elastic properties of pristine and defective *h*-BN. *Phys. Rev. B* **2017**, *96*, 184108. [CrossRef]

30. Thomas, S.; Ajith, K.M.; Chandra, S.; Valsakumar, M.C. Temperature dependent structural properties and bending rigidity of pristine and defective hexagonal boron nitride. *J. Phys. Condens. Matter* **2015**, *27*, 315302. [CrossRef] [PubMed]

31. Boldrin, L.; Scarpa, F.; Chowdhury, R.; Adhikari, S. Effective mechanical properties of hexagonal boron nitride nanosheets. *Nanotechnology* **2011**, *22*, 50. [CrossRef] [PubMed]

32. Genoese, A.; Genoese, A.; Rizzi, N.L.; Salerno, G. Force constants of BN, SiC, AlN and GaN sheets through discrete homogenization. *Meccanica* **2018**, *53*, 593–611. [CrossRef]

33. Thomas, S.; Ajith, K.M.; Valsakumar, M.C. Directional anisotropy, finite size effect and elastic properties of hexagonal boron nitride. *J. Phys. Condens. Matter* **2016**, *28*, 295302. [CrossRef] [PubMed]

34. Jiang, L.; Guo, W. A molecular mechanics study on size-dependent elastic properties of single-walled boron nitride nanotubes. *J. Mech. Phys. Solids* **2011**, *59*, 1204–1213. [CrossRef]

35. Kudin, K.N.; Scuseria, G.E.; Yakobson, B.I. C$_2$F, BN, and C nanoshell elasticity from ab initio computations. *Phys. Rev. B Condens. Matter Mater. Phys.* **2001**, *64*, 235406. [CrossRef]
36. Peng, Q.; Ji, W.; De, S. Mechanical properties of the hexagonal boron nitride monolayer: Ab initio study. *Comput. Mater. Sci.* **2012**, *56*, 11–17. [CrossRef]
37. Verma, V.; Jindal, V.K.; Dharamvir, K. Elastic moduli of a boron nitride nanotube. *Nanotechnology* **2007**, *18*, 43. [CrossRef]
38. Zhao, S.; Xue, J. Mechanical properties of hybrid graphene and hexagonal boron nitride sheets as revealed by molecular dynamic simulations. *J. Phys. D Appl. Phys.* **2013**, *46*, 13. [CrossRef]
39. Le, M.Q.; Nguyen, D.T. Atomistic simulations of pristine and defective hexagonal BN and SiC sheets under uniaxial tension. *Mater. Sci. Eng. A* **2014**, *615*, 481–488. [CrossRef]
40. Vodenitcharova, T.; Zhang, L.C. Effective wall thickness of a single-walled carbon nanotube. *Phys. Rev. B* **2003**, *68*, 165401. [CrossRef]
41. Wang, C.Y.; Zhang, L.C. A critical assessment of the elastic properties and effective wall thickness of single-walled carbon nanotubes. *Nanotechnology* **2008**, *19*, 7. [CrossRef] [PubMed]
42. Plimpton, S. Fast parallel algorithms for short-range molecular dynamics. *J. Comput. Phys.* **1995**, *117*, 1–19. [CrossRef]
43. Tersoff, J. New empirical approach for the structure and energy of covalent systems. *Phys. Rev. B* **1988**, *37*, 6991–7000. [CrossRef]
44. Tersoff, J. Modeling solid-state chemistry: Interatomic potentials for multicomponent systems. *Phys. Rev. B* **1989**, *39*, 5566–5568. [CrossRef]
45. Kinacı, A.; Haskins, J.B.; Sevik, C.; Çağın, T. Thermal conductivity of BN-C nanostructures. *Phys. Rev. B* **2012**, *86*, 115410. [CrossRef]
46. Rajasekaran, G.; Kumar, R.; Parashar, A. Tersoff potential with improved accuracy for simulating graphene in molecular dynamics environment. *Mater. Res. Express* **2016**, *3*, 3. [CrossRef]
47. Mortazavi, B.; Cuniberti, G. Mechanical properties of polycrystalline boron-nitride nanosheets. *RSC Adv.* **2014**, *4*, 19137–19143. [CrossRef]
48. Qi-Lin, X.; Zhen-Huan, L.; Xiao-Geng, T. The defect-induced fracture behaviors of hexagonal boron-nitride monolayer nanosheets under uniaxial tension. *J. Phys. D Appl. Phys.* **2015**, *48*, 37. [CrossRef]
49. Wei, A.; Li, Y.; Datta, D.; Guo, H.; Lv, Z. Mechanical properties of graphene grain boundary and hexagonal boron nitride lateral heterostructure with controlled domain size. *Comput. Mater. Sci.* **2017**, *126*, 474–478. [CrossRef]
50. Tabarraei, A.; Wang, X. A molecular dynamics study of nanofracture in monolayer boron nitride. *Mater. Sci. Eng. A* **2015**, *641*, 225–230. [CrossRef]
51. Wong, C.H.; Vijayaraghavan, V. Nanomechanics of free form and water submerged single layer graphene sheet under axial tension by using molecular dynamics simulation. *Mater. Sci. Eng. A* **2012**, *556*, 420–428. [CrossRef]
52. Zhang, Y.Y.; Xiang, Y.; Wang, C.M. Buckling of defective carbon nanotubes. *J. Appl. Phys.* **2009**, *106*, 620–653. [CrossRef]
53. Yakobson, B.I.; Brabec, C.J.; Bernholc, J. Nanomechanics of carbon tubes: Instabilities beyond linear response. *Phys. Rev. Lett.* **1996**, *76*, 2511–2514. [CrossRef] [PubMed]
54. Huang, Y.; Wu, J.; Hwang, K.C. Thickness of graphene and single-wall carbon nanotubes. *Phys. Rev. B* **2006**, *74*, 245413. [CrossRef]
55. Batra, R.C.; Gupta, S.S. Wall thickness and radial breathing modes of single-walled carbon nanotubes. *J. Appl. Mech.* **2008**, *75*, 0610101–0610106. [CrossRef]

nanomaterials

MDPI

Article

Direct Observation of Inner-Layer Inward Contractions of Multiwalled Boron Nitride Nanotubes upon in Situ Heating

Zhongwen Li [1,2], Zi-An Li [1,*], Shuaishuai Sun [1], Dingguo Zheng [1,2], Hong Wang [1,2], Huanfang Tian [1], Huaixin Yang [1,2], Xuedong Bai [1,2,3] and Jianqi Li [1,2,3,*]

[1]   Beijing National Laboratory for Condensed Matter Physics, Institute of Physics, Chinese Academy of Sciences, Beijing 100190, China; lizhongwen@iphy.ac.cn (Z.L.); sss@iphy.ac.cn (S.S.); zdg@iphy.ac.cn (D.Z.); hong.w@iphy.ac.cn (H.W.); hftian@iphy.ac.cn (H.T.); hxyang@iphy.ac.cn (H.Y.); xdbai@iphy.ac.cn (X.B.)
[2]   School of Physical Sciences, University of Chinese Academy of Sciences, Beijing 100049, China
[3]   Collaborative Innovation Center of Quantum Matter, Beijing 100084, China
*   Correspondence: zali79@iphy.ac.cn (Z.-A.L.); ljq@iphy.ac.cn (J.L.); Tel.: +86-10-82649524 (J.L.)

Received: 18 January 2018; Accepted: 30 January 2018; Published: 4 February 2018

Abstract: In situ heating transmission electron microscopy observations clearly reveal remarkable interlayer expansion and inner-layer inward contraction in multi-walled boron nitride nanotubes (BNNTs) as the specimen temperature increases. We interpreted the observed inward contraction as being due to the presence of the strong constraints of the outer layers on radial expansion in the tubular structure upon in situ heating. The increase in specimen temperature upon heating can create pressure and stress toward the tubular center, which drive the lattice motion and yield inner diameter contraction for the multi-walled BNNTs. Using a simple model involving a wave-like pattern of layer-wise distortion, we discuss these peculiar structural alterations and the anisotropic thermal expansion properties of the tubular structures. Moreover, our in situ atomic images also reveal Russian-doll-type BN nanotubes, which show anisotropic thermal expansion behaviors.

Keywords: multi-walled BNNTs; anisotropic thermal expansion; transmission electron microscopy; in situ heating; thermal contraction

1. Introduction

In the recent decades, one-dimensional (1D) tubular nanomaterials, notably the carbon-based nanotubes and the mineral-based nanotubes [1–3], have attracted a great deal of attention for many applications because of their novel physical properties [4–6]. Carbon nanotubes (CNTs) are one prominent example, and exhibit rich transport properties including insulating, semiconducting and metallic behaviors, depending on the tubular chirality [7]. Boron nitride nanotubes (BNNTs), structurally similar to CNTs, also have tubular structures, but only exhibit semiconducting transport with a large band gap, regardless of tubular chirality [8,9]. BNNTs also exhibit high chemical stability, excellent mechanical properties, and high thermal conductivity. These unique properties make BN nanotubes a promising nanomaterial in a variety of potential fields, such as gas sensing, spin filters, bio sensing, functional composites, hydrogen accumulators, and electrically insulating substrates [9–13]. Exploitation of nanoscale electronic devices based on these tubular nanostructures requires a full understanding of not only the static structural characteristics, but also the structural response to externally applied stimuli [14], e.g., thermal heat and electromagnetic waves. However, the nanosized tubular structure imposes great challenges in investigating their structural response on the individual nanotube level.

High-resolution transmission electron microscopy (HRTEM) allows a direct imaging of the atomic column of thin specimens, and has become an indispensable tool for studying nanosized materials at an atomic resolution. In recent years, in situ experimentation in HRTEM has emerged as an exciting field of study [15]. The application of HRTEM to studying dynamic changes in a specimen at atomic scale have met with challenges due to the lack of mechanical stability of specimens when applying stimuli, e.g., in situ heating. Recently, with the advancements in microelectromechanical systems (MEMS)-based technologies, in situ HRTEM studies have progressed rapidly over the last decade. Here, we employ the MEMS-based in situ heating holder to study the thermal properties of BNNTs, with a particular focus on the atomic-resolution layer-by-layer lattice change due to in situ heating. We aim to understand the peculiar thermal expansion in the interlayer of van der Waals bonds and the thermal contraction in the intralayer of strong covalent bonds in the tubular BNNTs.

2. Experimental

The in situ heating TEM observations were performed on a Jeol-2100F TEM (JEOL Ltd., Tokyo, Japan) operating under an acceleration voltage of 200 kV. A chip-type heating TEM holder (Protochip, Inc., Morrisville, NC, USA) was used to obtain the electron diffraction and high-resolution lattice images at high temperatures. Our samples of BNNTs were synthesized by Advanced Materials Laboratory, National Institute for Materials Science (NIMS), Japan. More information about the samples can be obtained from ref. [16,17]. The multi-walled BN nanotubes were dispersed in ethanol using an ultrasonicator for half an hour. Then the mixed liquid was dropped on the micro-grid of the heating holder. The local sample temperature can be varied from room temperature to 1200 K. For accurate lattice measurement with high precision, the TEM imaging and diffraction conditions and the specimen positions are kept nearly identical at different specimen temperatures.

3. Results and Discussion

Prior to discussing the experimental results, we will firstly discuss the three tubular atomic models of BN nanotube in the literature. It is commonly known that multi-walled nanotubes have three types of tubular structures, i.e., Russian doll, jelly scroll, and mixture structures [18], as shown in Figure 1a. They have all been directly observed in transmission electron microscope (TEM) investigations [19]. Measurements of thermal expansions by means of X-ray and electron scatterings have clearly demonstrated that the interlayer spacings between atomic sheets often increase evidently from (l) to ($l + \Delta l$) upon the increase of temperature due to the anharmonic nature of inter-atomic Van der Waals interaction.

For the doll model, we can take the armchair nanotube as an example, which consists of C1 (circumferential) and C2 (axial) bonds [20], as shown in Figure 1b. If we only consider the variation of nanotubes caused by the alterations of chemical bonds during temperature rise, the variation of axial length should be comparable with that along the radial direction, which can be estimated by ($1 + \Delta l/l$) times the original length. For the Russian doll model, as shown in Figure 1b, the thermal expansion in the axial and radial directions should adopt a similar feature. For the Jelly scroll and the mixture models, anisotropic lattice expansions can be observed, and visibly large lattice expansion occurs in the radial direction due to the weak Van der Waals interaction between layers. On the other hand, based on a large number of results from XRD and TEM experimental observations, nanotubes often show notable anisotropic characteristics during thermal expansion. For example, the thermal expansion coefficient of carbon nanotubes in the radial direction is 2.5×10^{-5} K^{-1}, while the lattice expansion in the axial direction is only 1/10 of the former [18,21]. Figure 1c shows the temperature-dependent lattice changes in the axial and radial directions obtained from a series of electron diffraction patterns from 300 to 1100 K by in situ heating of BNNTs inside the TEM. The BNNTs exhibit a notable linear expansion in the radial direction, and the coefficient of expansion is measured to be 3.5×10^{-5} K^{-1}. By contrast, the axial direction exhibits a negative thermal expansion behavior, which is fitted with a two-term polynomial function [22]. These data are comparable with those obtained from graphite and H-BN crystals [23]. Therefore, the main experimental results, in particular the large inter-sheet thermo-expansion, are interpreted to be based on Jelly scroll and mixture models [18,24,25]. However, according to our recent study on the microstructures of a number of well-characterized samples, a large fraction of the BNNTs adopt the Russian doll structure. Moreover, these multi-walled nanotubes exhibit visible inner diameter contraction associated with lattice expansion as temperature increases.

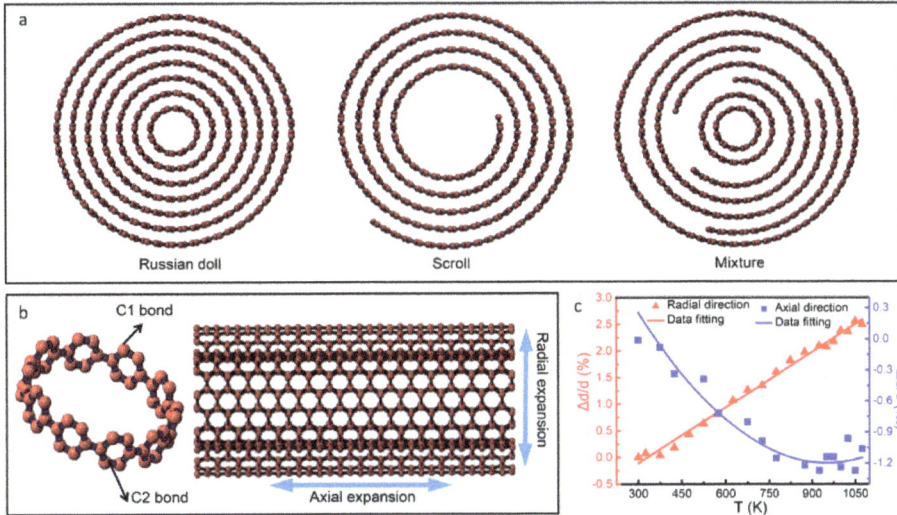

Figure 1. Atomic models for tubular structure, the radial and axial directions are indicated. (**a**) Three types of BN tubular structure: Russian doll, scroll and the mixture; (**b**) Lattice change in the radial and axial directions in nanotubes of Russian doll; (**c**) Thermal coefficient of multi-walled boron nitride nanotubes using the in situ TEM heating experiment.

Figure 2 shows the atomic-resolved lattice images obtained as a function of applied temperature, from which the interlayer spacing and the diameter of the innermost layer are measured [26]. Figure 2a shows a typical high-resolution image of BNNTs at room temperature, in which the lattice spacing, as well the innermost layer diameters for the measured tubular, can be simultaneously obtained. Figure 2b shows the line profile traces of the high-resolution lattice image in the direction perpendicular to the nanotube axis (boxed area). To improve the lattice measurement accuracy, the broad width of the scan line is used, as indicated by the boxed area. The periodical line profiles trace the tubular layers and interlayer spacing of the multi-walled BNNTs. Figure 2c illustrates the measurement of the diameter of the innermost layer of BNNTs, as measured directly between the two innermost layers. We note that in situ heating of the specimen with varying temperature leads to a slight specimen drifts. The high-resolution images were taken after the pre-set temperature was reached and the specimen was found to be stabilized. We use the specific structural features in the nanotube as a reference mark, and the lattice images of different temperature were realigned prior to lattice measurement. In order to reduce the degree of experimental error, we used average data taken from a series of atomic-resolved images at the same temperature for the data processing.

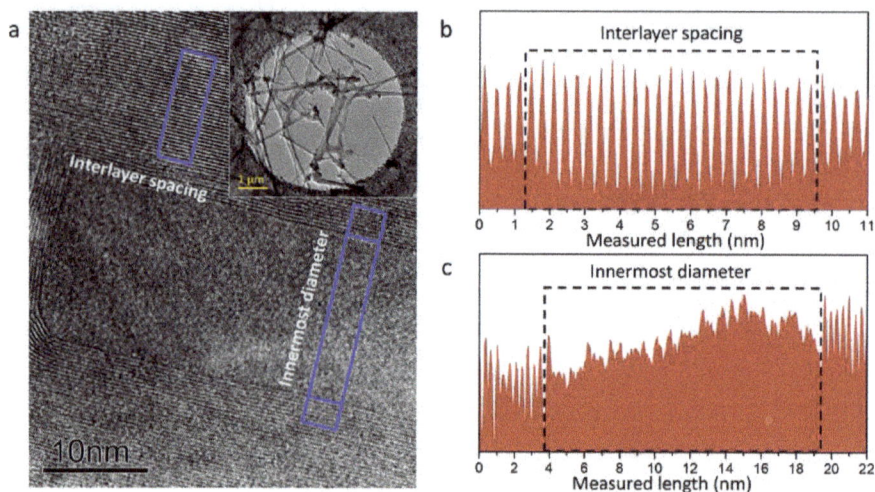

Figure 2. High-resolution lattice image of multi-walled boron nitride nanotubes BNNTs. (**a**) High-resolution lattice images of BNNTs, in which the interlayer spacing and the diameter of the innermost layer are indicated. The inset shows a low-magnification TEM image of the BNNT assembly; Line-scan profiles trace the interlayer spacing in (**b**), and the diameter of the innermost layer in (**c**).

Three sets of in situ heating experimental results from room temperature up to 1100 K are shown in Figure 3. Figure 3a shows in situ heating experimental images for a multi-walled BNNT. Using the line-scan profile method as described in Figure 2, we measured the interlayer spacing and the diameter of the innermost layer from Figure 3a, as shown in Figure 3b. It is recognizable that the innermost diameter decreases progressively with the temperature rise. In contrast, the interlayer spacings become larger with temperature rise. Importantly, the experimental results for layer spacing modification are basically consistent with the data calculated by using thermal expansion with the coefficient as discussed in the above context. This fact suggests that the in situ heating TEM images reliably reveal the lattice expansion in the multi-walled tubular structure. At a temperature of 1100 K, the experimental data for the inner diameter of this nanotube contracts by 1.1 nm (11%) in comparison with the room temperature data.

Figure 3c shows a second sample with an additional cap-like tubular structure. Figure 3d presents the interlayer spacing and innermost diameter as a function of temperature. It is notable that the second nanotube has a closed cap-like structure in the inner layer, which actually features the multi-walled BN nanotube for the Russian doll structure, as reported for BN tubes in the microstructural analysis. The inner diameter decreases as temperature increases, and the lattice distance between interlayers increases as temperature increases. The changes of the interlayer spacing of this nanotube show a fundamentally similar feature with the previous one. However, the contraction of the innermost diameter (~3%) is much smaller than that (11%) of the first one (Figure 3a,b). Obviously, the cap-like tubular structure inside the tube of the second sample (see Figure 3c) play a role in hindering the inward contraction of the nanotube.

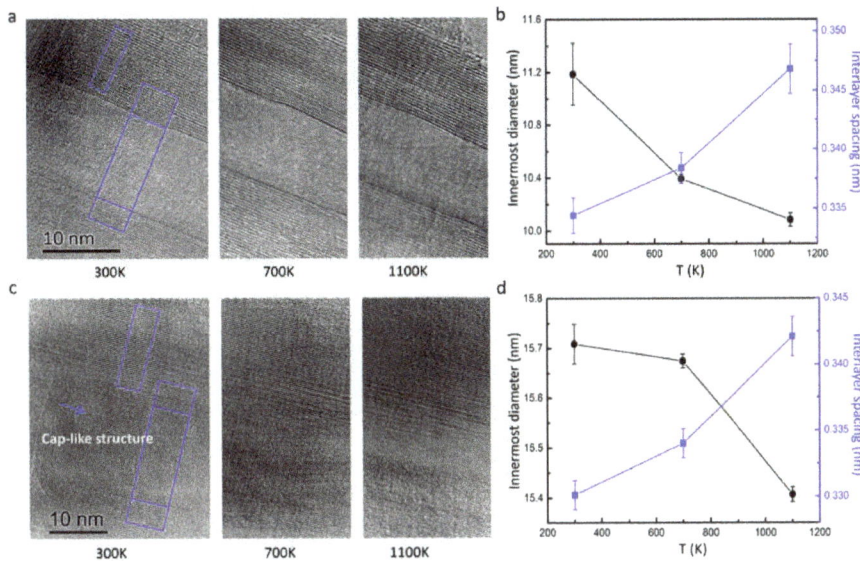

Figure 3. High-resolution lattice images of BNNTs from in situ TEM heating experiments. (**a**) High-resolution lattice images of BNNTs taken at 300, 700 and 1100 K, respectively; (**b**) the corresponding lattice spacings of the axial and radial directions for the three temperatures; (**c,d**) are the same as (**a,b**), but for nanotubes with a cap-like structures.

It is known that the interlayer spacing of multi-walled BN nanotubes depends mainly on the weak Van der Waals interaction, as extensively discussed in the previous literature. This can be greatly affected by the morphologic features of specific BN sheets in nano-tubular structures [27,28]. In the present case, we suggest that the strong constraints exerted by the outer layers prevent the lattice expansion of multi-walled BNNTs in the radially outward direction. Conversely, the temperature rise could create pressure and stress on the inner layers of the tubes, and this type of stress could release progressively with the BN inner-layer motions radially inward to the tubular center, i.e., the contraction of the innermost layers associated with the increase in temperature, as seen in our in situ heating TEM observations. In fact, similar experimental phenomena have been noted in metal-filled multi-walled carbon nanotubes. Sun and coworkers have reported that, with increasing specimen temperature, the metal or carbide nanowires inside the carbon nanotube are subjected to an inner pressure of up to 40 GPa because of the contraction of inner tubular layers [29].

Admittedly, using static high-resolution atomically resolved imaging in the present study, it is still difficult to obtain a microscopic understanding of the structural distortion and lattice motions occurring in the inner tubular layers following in situ heating. The contraction in axial direction of the BN nanotubes (Figure 1c) is generally believed to be caused by certain anharmonic phonon modes of the multi-walled nanotubes [30]. Therefore, we turn to a descriptive model discussion based on the morphology alterations associated with thermo-excitation of phonon modes in tubular structure [31]. The phonon modes of nanotubes can be categorized into three types: torsional, longitudinal and radial modes. Because the interlayer Van der Waals interaction is determined by the radial direction only, it has almost no effect on torsional and longitudinal mode [32,33].

The radial modes have major effects, and it is shown that the radial mode can cause the contraction of the inner diameter of the nanotubes, whereby the atoms on the cross-section of the nanotubes can be defined as follows (Figure 4a),

$$r_1 = r(T) + A \cdot \sin(\omega \cdot \theta) \tag{1}$$

where r_1 is the distance from each atom to the center of the cross-section, $r(T)$ is the average distance from each atom to the center at a certain temperature T, and A is the amplitude of displacement, and $1/\omega$ is wavelength of the phonon mode. Obviously, the inner diameter of the nanotube exhibits clear contraction in comparison with the outer layers. We then do the following calculations,

$$2\pi r_0 = \int_0^{2\pi} \sqrt{dr_1^2 + (r_1 \cdot d\theta)^2}$$

(2)

where r_0 is the radius of the nanotube at 0 K. In our experiments, r_1 is much larger than A, so the following results can be obtained when Equation (2) is expanded by the Taylor formula, neglecting the higher-order term.

$$r_0 - r(T) = \frac{(2\omega^2 - 1)A^2}{r_0}$$

(3)

where $r_0 - r(T)$ is the contraction, Δr, of the nanotube relative to 0 K. As a result, the smaller the inner diameter of the monolayer multi-wall nanotubes is, the greater the shrinkage that occurs, as shown in Figure 4b. Generally speaking, with a rise in temperature, there is no new vibration mode or frequency activated; instead, only the vibration amplitude increases [34]. The interlayer spacing of the nanotubes will become larger due to the shrinkage of the inner layer driven by increasing specimen temperature.

The A^2 can be considered to be linearly increasing upon temperature rise. When the specimen temperature increases, the inner-layer diameter of the tubular structure tends to shrink, and the interlayer spacing will increase accordingly. We therefore suggest that the structural evolutions of the multi-walled BN nanotubes evidently depend on both the inter-layer bonds and the strong restriction arising initially from the outer tubular layers, which explains well the anisotropic thermal expansion of the nanotubes observed in our in situ heating experiment.

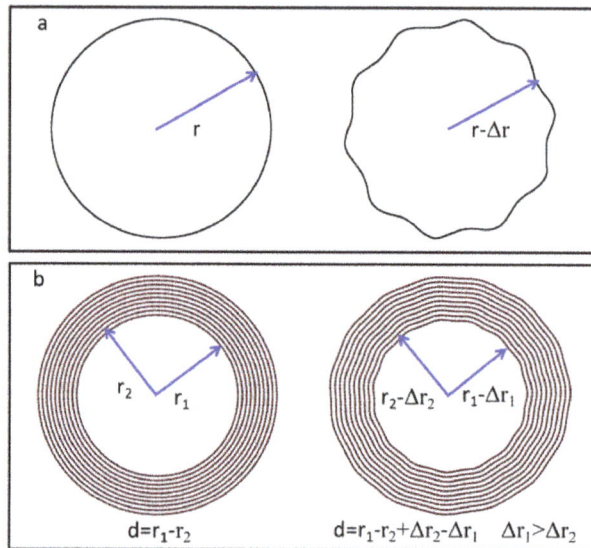

Figure 4. Schematic models for the contraction of innermost layer of nanotube due to the heating effect on the tubular structure (in the cross-section view). (**a**) The left-panel shows the schematic diagram without the wave-like atomic displacement, and the right-panel shows the lattice contraction caused by the wave-like atomic displacement; (**b**) Illustrative schematics of lattice contraction of innermost layers of multi-walled tubular structure upon in situ heating.

4. Conclusions

Using in situ high-resolution TEM and the MEMS-based heating holder, we have studied the thermal lattice change in the multi-walled boron nitride nanotubes (BNNTs). Analysis of the atomically resolved lattice images at different applied temperature reveals the inner-layer contraction of the BN nanotube with an increase in specimen temperature. To understand the inner-layer inward contraction, we performed a simple model calculation to illustrate the structural alterations associated with the thermal expansion in multi-walled boron nitride tubes of doll model. Based on the model calculation, we interpret the observed inward contraction as being due to the presence of strong constraints of the outer-layers on the radial expansion in the tubular structure upon in situ heating.

Acknowledgments: This work was supported by the National Key Research and Development Program of China (Nos. 2016YFA0300303, 2017YFA0504703, and 2017YFA0302900), the National Basic Research Program of China 973 Program (No. 2015CB921300), the Natural Science Foundation of China (Nos. 11447703, 11604372, 11474323, 11774403 and 11774391), "Strategic Priority Research Program (B)" of the Chinese Academy of Sciences (No. XDB07020000), and the Scientific Instrument Developing Project of the Chinese Academy of Sciences (No. ZDKYYQ20170002). Z.A.L acknowledges the financial support from the CAS Pioneer Hundred Talents Program B.

Author Contributions: J.L. and Z.-A.L. conceived and designed the experiments; Z.L., S.S, D.Z, and H.W. performed the experiments; Z.L., S.S., H.T., H.Y. analyzed the data; X.B. contributed to the BN materials; Z.L. wrote the paper with contributions from all co-authors.

Conflicts of Interest: The authors declare no conflict of interest.

References

1. Cavallaro, G.; Danilushkina, A.A.; Evtugyn, V.G.; Lazzara, G.; Milioto, S.; Parisi, F.; Rozhina, E.V.; Fakhrullin, R.F. Halloysite Nanotubes: Controlled Access and Release by Smart Gates. *Nanomaterials* **2017**, *7*, 199. [CrossRef] [PubMed]

2. Yang, Y.; Chen, Y.; Leng, F.; Huang, L.; Wang, Z.; Tian, W. Recent Advances on Surface Modification of Halloysite Nanotubes for Multifunctional Applications. *Appl. Sci.* **2017**, *7*, 1215. [CrossRef]

3. Makaremi, M.; Pasbakhsh, P.; Cavallaro, G.; Lazzara, G.; Aw, Y.K.; Lee, S.M.; Milioto, S. Effect of Morphology and Size of Halloysite Nanotubes on Functional Pectin Bionanocomposites for Food Packaging Applications. *ACS Appl. Mater. Interfaces* **2017**, *9*, 17476–17488. [CrossRef] [PubMed]

4. Mceuen, P.L. Nanotechnology: Carbon-based electronics. *Nature* **1998**, *393*, 15–17. [CrossRef]

5. Lu, W.; Lieber, C.M. Nanoelectronics from the bottom up. *Nat. Mater.* **2007**, *6*, 841–850. [CrossRef] [PubMed]

6. Lee, C.H.; Qin, S.; Savaikar, M.A.; Wang, J.; Hao, B.; Zhang, D.; Banyai, D.; Jaszczak, J.A.; Clark, K.W.; Idrobo, J.C.; et al. Room-temperature tunneling behavior of boron nitride nanotubes functionalized with gold quantum dots. *Adv. Mater.* **2013**, *25*, 4544–4548. [CrossRef] [PubMed]

7. Anantram, M.P.; Léonard, F. Physics of carbon nanotube electronic devices. *Rep. Prog. Phys.* **2006**, *69*, 507–561. [CrossRef]

8. Golberg, D.; Bando, Y.; Tang, C.C.; Zhi, C.Y. Boron Nitride Nanotubes. *Adv. Mater.* **2007**, *19*, 2413–2432. [CrossRef]

9. Golberg, D.; Bando, Y.; Huang, Y.; Terao, T.; Mitome, M.; Tang, C.; Zhi, C. Boron Nitride Nanotubes and Nanosheets. *ACS Nano* **2010**, *4*, 2979–2993. [CrossRef] [PubMed]

10. Xiang, C.; Chen, T.; Zhang, H.; Zou, Y.; Chu, H.; Zhang, H.; Xu, F.; Sun, L.; Tang, C. Growth of copper-benzene-1,3,5-tricarboxylate on boron nitride nanotubes and application of the composite in methane sensing. *Appl. Surf. Sci.* **2017**, *424*, 39–44. [CrossRef]

11. Dhungana, K.B.; Pati, R. Boron nitride nanotubes for spintronics. *Sensors* **2014**, *14*, 17655–17685. [CrossRef] [PubMed]

12. Farmanzadeh, D.; Ghazanfary, S. Interaction of vitamins B_3 and C and their radicals with (5, 0) single-walled boron nitride nanotube for use as biosensor or in drug delivery. *J. Chem. Sci.* **2013**, *125*, 1595–1606. [CrossRef]

13. Chen, X.; Wu, P.; Rousseas, M.; Okawa, D.; Gartner, Z.; Zettl, A.; Bertozzi, C.R. Boron Nitride Nanotubes Are Noncytotoxic and Can Be Functionalized for Interaction with Proteins and Cells. *J. Am. Chem. Soc.* **2009**, *131*, 890–891. [CrossRef] [PubMed]

14. Li, D.; Paxton, W.F.; Baughman, R.H.; Huang, T.J.; Stoddart, J.F.; Weiss, P.S. Molecular, Supramolecular, and Macromolecular Motors and Artificial Muscles. *MRS Bull.* **2011**, *34*, 671–681. [CrossRef]
15. Ramachandramoorthy, R.; Bernal, R.; Espinosa, H.D. Pushing the envelope of in situ transmission electron microscopy. *ACS Nano* **2015**, *9*, 4675–4685. [CrossRef] [PubMed]
16. Golberg, D.; Bai, X.D.; Mitome, M.; Tang, C.C.; Zhi, C.Y.; Bando, Y. Structural peculiarities of in situ deformation of a multi-walled BN nanotube inside a high-resolution analytical transmission electron microscope. *Acta Mater.* **2007**, *55*, 1293–1298. [CrossRef]
17. Zhi, C.; Bando, Y.; Tan, C.; Golberg, D. Effective precursor for high yield synthesis of pure BN nanotubes. *Solid State Commun.* **2005**, *135*, 67–70. [CrossRef]
18. Maniwa, Y.; Fujiwara, R.; Kira, H.; Tou, H.; Nishibori, E.; Takata, M.; Sakata, M.; Fujiwara, A.; Zhao, X.; Iijima, S.; et al. Multiwalled carbon nanotubes grown in hydrogen atmosphere: An X-ray diffraction study. *Phys. Rev. B* **2001**, *64*, 073105. [CrossRef]
19. Lavin, J.G.; Subramoney, S.; Ruoff, R.S.; Berber, S.; Tománek, D. Scrolls and nested tubes in multiwall carbon nanotubes. *Carbon* **2002**, *40*, 1123–1130. [CrossRef]
20. Cao, G.; Chen, X.; Kysar, J.W. Apparent thermal contraction of single-walled carbon nanotubes. *Phys. Rev. B* **2005**, *72*, 235404. [CrossRef]
21. Bandow, S. Radial Thermal Expansion of Purified Multiwall Carbon Nanotubes Measured by X-ray Diffraction. *Jpn. J. Appl. Phys.* **1997**, *36*, L1403–L1405. [CrossRef]
22. Yates, B.; Overy, M.J.; Pirgon, O. The anisotropic thermal expansion of boron nitride. *Philos. Mag.* **2006**, *32*, 847–857. [CrossRef]
23. Tohei, T.; Lee, H.-S.; Ikuhara, Y. First Principles Calculation of Thermal Expansion of Carbon and Boron Nitrides Based on Quasi-Harmonic Approximation. *Mater. Trans.* **2015**, *56*, 1452–1456. [CrossRef]
24. Park, S.T.; Flannigan, D.J.; Zewail, A.H. 4D electron microscopy visualization of anisotropic atomic motions in carbon nanotubes. *J. Am. Chem. Soc.* **2012**, *134*, 9146–9149. [CrossRef] [PubMed]
25. Cao, G.; Sun, S.; Li, Z.; Tian, H.; Yang, H.; Li, J. Clocking the anisotropic lattice dynamics of multi-walled carbon nanotubes by four-dimensional ultrafast transmission electron microscopy. *Sci. Rep.* **2015**, *5*, 8404. [CrossRef] [PubMed]
26. Kiang, C.H.; Endo, M.; Ajayan, P.M.; Dresselhaus, G.; Dresselhaus, M.S. Size Effects in Carbon Nanotubes. *Phys. Rev. Lett.* **1998**, *81*, 1869–1872. [CrossRef]
27. Salehi-Khojin, A.; Jalili, N. Buckling of boron nitride nanotube reinforced piezoelectric polymeric composites subject to combined electro-thermo-mechanical loadings. *Compos. Sci. Technol.* **2008**, *68*, 1489–1501. [CrossRef]
28. Thomas, S.; Ajith, K.M.; Chandra, S.; Valsakumar, M.C. Temperature dependent structural properties and bending rigidity of pristine and defective hexagonal boron nitride. *J. Phys. Condens. Matter Inst. Phys. J.* **2015**, *27*, 315302. [CrossRef] [PubMed]
29. Sun, J.; Xu, F.; Sun, L.-T. In situ investigation of the mechanical properties of nanomaterials by transmission electron microscopy. *Acta Mech. Sin.* **2012**, *28*, 1513–1527. [CrossRef]
30. Mashreghi, A. Thermal expansion/contraction of boron nitride nanotubes in axial, radial and circumferential directions. *Comput. Mater. Sci.* **2012**, *65*, 356–364. [CrossRef]
31. Kwon, Y.K.; Berber, S.; Tomanek, D. Thermal contraction of carbon fullerenes and nanotubes. *Phys. Rev. Lett.* **2004**, *92*, 015901. [CrossRef] [PubMed]
32. He, X.Q.; Eisenberger, M.; Liew, K.M. The effect of van der Waals interaction modeling on the vibration characteristics of multiwalled carbon nanotubes. *J. Appl. Phys.* **2006**, *100*, 124317. [CrossRef]
33. Wang, C.Y.; Ru, C.Q.; Mioduchowski, A. Free vibration of multiwall carbon nanotubes. *J. Appl. Phys.* **2005**, *97*, 114323. [CrossRef]
34. Cao, G.; Chen, X.; Kysar, J.W. Thermal vibration and apparent thermal contraction of single-walled carbon nanotubes. *J. Mech. Phys. Solids* **2006**, *54*, 1206–1236. [CrossRef]

nanomaterials

MDPI

Article

Alignment of Boron Nitride Nanofibers in Epoxy Composite Films for Thermal Conductivity and Dielectric Breakdown Strength Improvement

Zhengdong Wang [1,†]**, Jingya Liu** [1,†]**, Yonghong Cheng** [1,*]**, Siyu Chen** [1]**, Mengmeng Yang** [1]**, Jialiang Huang** [1]**, Hongkang Wang** [1]**, Guanglei Wu** [1,2] **and Hongjing Wu** [3]

[1] Center of Nanomaterials for Renewable Energy, State Key Laboratory of Electrical Insulation and Power Equipment, Xi'an Jiaotong University, Xi'an 710049, China; zhengdong.wang@stu.xjtu.edu.cn (Z.W.); ljy950417@stu.xjtu.edu.cn (J.L.); chensiyu0901@stu.xjtu.edu.cn (S.C.); belleyoung@stu.xjtu.edu.cn (M.Y.); huangjialiang@stu.xjtu.edu.cn (J.H.); hongkang.wang@mail.xjtu.edu.cn (H.W.); wuguanglei@mail.xjtu.edu.cn (G.W.)

[2] Institute of Materials for Energy and Environment, Growing Base for State Key Laboratory, College of Materials Science and Engineering, Qingdao University, Qingdao 266071, China

[3] Department of Applied Physics, Northwestern Polytechnical University, Xi'an, 710072, China; wuhongjing@mail.nwpu.edu.cn

* Correspondence: cyh@mail.xjtu.edu.cn; Tel.: +86-029-82665178

† These authors contributed equally to this work.

Received: 27 March 2018; Accepted: 10 April 2018; Published: 15 April 2018

Abstract: Development of polymer-based composites with simultaneously high thermal conductivity and breakdown strength has attracted considerable attention owing to their important applications in both electronic and electric industries. In this work, boron nitride (BN) nanofibers (BNNF) are successfully prepared as fillers, which are used for epoxy composites. In addition, the BNNF in epoxy composites are aligned by using a film casting method. The composites show enhanced thermal conductivity and dielectric breakdown strength. For instance, after doping with BNNF of 2 wt%, the thermal conductivity of composites increased by 36.4% in comparison with that of the epoxy matrix. Meanwhile, the breakdown strength of the composite with 1 wt% BNNF is 122.9 kV/mm, which increased by 6.8% more than that of neat epoxy (115.1 kV/mm). Moreover, the composites have maintained a low dielectric constant and alternating current conductivity among the range of full frequency, and show a higher thermal decomposition temperature and glass-transition temperature. The composites with aligning BNNF have wide application prospects in electronic packaging material and printed circuit boards.

Keywords: electrospinning technique; BN nanofibers; alignment; epoxy composite; thermal and dielectric properties

1. Introduction

In recent years, to achieve a better performance, the tendency in electronic technology has been towards integrating more transistors into a single chip. Yet this has caused the escalation of energy dissipation and significant heat fluxes in a device, which negatively influences its lifetime and reliability [1–5]. So, novel materials with better electrical and thermal performance are needed. Epoxy resin is one of the most frequently used insulation materials, with superior electrical and machining properties [6,7]. However, its low thermal conductivity greatly restricts its applications because of the increasing heat produced by unit volume of equipment [8,9]. To improve thermal conductivity of epoxy resin, the fillers of micro-scale with high thermal conductivity can be considered. Nevertheless, the dilemma for using micro-sized inorganic particles is that the increase of thermal conductivity is usually attained at the expense of a significant decrease in breakdown strength, which thus limits their applications in high output electrical devices [2,10,11]. Thus, it can be seen that the development of epoxy-based composites with high thermal conductivity and dielectric breakdown strength is impending.

In the field of electrical insulation, inorganic particles with high intrinsic thermal conductivity and desirable insulating properties are considered more as fillers of composites, such as alumina (Al_2O_3), aluminum nitride (AlN) and boron nitride (BN) [2,8,12–14]. Among these fillers, boron nitride has become a research focus because of its high thermal conductivity (intrinsic ~2000 W/m K), low permittivity ε' ~4, loss and density. In addition, for the high aspect ratio, advances are focused on one-dimensional boron nitride nanotubes (BNNTs) and two-dimensional boron nitride nanosheets (BNNSs) [1,15,16].

For 2D thermal transport materials, graphene polymer composites exhibit high thermal transport and high electric conductivity, exhibiting many applications in the field of electronics [17–19]. BNNSs are very promising materials for application in the electrical insulating field. But there are still some problems and bottlenecks that need to be addressed. Different from graphene, BNNSs are brittle and have a stronger interlayer interaction, making them difficult to exfoliate. For 1D materials, similar to carbon nanotubes (CNTs), boron nitride nanotubes (BNNTs) have a single-walled or multi-walled tubular hexagonal structure [20–22]. It is worth noting that BNNTs have a large band gap and superior thermal transport properties while CNTs are conductive. Hence, BNNTs have great potential for high thermal conductivity and insulating composite materials [1,23]. However, the preparation methods for BNNT, such as the arc-melting method and Chemical Vapor Deposition, are complex to operate [24–26]. Moreover, the yield of BNNTs is low and the control of the perfect structure is fairly difficult, limiting its practical application for significantly improving thermal conductivity.

In this work, the electrospinning method is applied to prepare the nanofibers. Recently, the electrospinning method has attracted more and more attention due to its many advantages, such as easy preparation technology, high-yield nanomaterials and regular structure of nanofibers for energy storage and thermal management materials [27–30]. In addition, nanofibers prepared by electrospinning have a large aspect ratio because of their long length and controllable diameter, which is beneficial for forming a connected network for the improvement of thermal conductivity of nanocomposites [31–33]. Moreover, dielectric fillers with a high aspect ratio are found to be effective in improving breakdown strength [23]. Furthermore, it was found to be possible that breakdown strength may be improved as the nanofibers oriented preferentially in the in-plane directions of the composite films [34]. In order to realize the orientation of the nanofibers, the coating method would be applied. In this paper, the orientation-BN nanofiber/Epoxy composite film with increasing breakdown strength and thermal conductivity is developed. Utilizing the electrospinning method to fabricate the BN nanofibers and applying the coating method to orient the nanofibers inside the composite film. The electrical performance of the products will be evaluated including dielectric properties and the electrical breakdown strength at a power frequency alternating current (AC) system; On the other hand, BNNF incorporated into polymer can enhance the thermal-mechanical dielectric properties of polymer composites by restraining the free mobility of polymer molecular chains, such as thermal stability and glass-transition temperature, which are also crucial for packaging materials in powerful micro-electronic devices [35–39]. The thermal performance of the products has been evaluated, including thermal conductivity, thermal gravimetric analysis (TGA) and differential scanning calorimetry (DSC).

2. Experimental

2.1. Material

Polyvinyl butyral (PVB) powder was bought from Sinopharm Chemical Reagent (Shanghai) Co. Ltd., China. Boric oxide (B_2O_3) was purchased from Aladdin Reagent (Shanghai) Co. Ltd., China.

Absolute ethyl alcohol was purchased from Zhiyuan Chemical Reagent (Tianjin) Co. Ltd., China. The bisphenol A epoxy resin (E-828) along with anhydrite hardener (MTHPA) and benzyldimethylamine (BDMA, used as the accelerant) were purchased from Aladdin Reagent (Shanghai) Co. Ltd., China.

2.2. Preparation of BNNF

As shown in Figure 1, first, 1.4 g of boron oxide powder was dissolved in 20 mL of absolute ethanol solvent and stirred at 45°C with a magnetic stirrer until completely melted. Then, 0.7 g of PVB (M_w = 60,000) powder was dissolved in 20 mL of absolute ethanol solvent and poured into B_2O_3/ethanol solvent. The mixture was stirred vigorously at 45 °C for 24 h. Next, the electrospinning method was used to prepare BN nanofibers. The output voltage of electrospinning was 16 kV. Besides, to prevent occluding the needle with precipitate from solvent evaporation during the process of electrospinning, the syringe propeller was applied to control the flow rate of the jet flow. The prepared solution was injected via a syringe with a feeding rate of 1.5 mL/h. Then, the PVB is removed from the B_2O_3/PVB fiber (BOF) and the BOF was nitrided. The BOF was put into a crucible, which was large enough to make the fiber stretch; the crucible was put into a sintering furnace with a 10% O_2/90% NH_3 (vol/vol) gaseous mixture with a total flux of 200 mL/min, while at the same time, heating it from room temperature to 800 °C. After 2 h of reaction under 800 °C, the polymer from the BOF was removed. The gaseous mixture was converted into NH_3 (99.99%) at 100 mL/min, heating from 800 °C to 1100 °C and then held at 1100 °C for nitriding for 6 h. Secondly, the atmosphere changed into N_2 (99.99%) at 200 mL/min, heating from 1100 °C to 1500 °C with a rate of 1 K/min and was then held at 1500 °C for nitriding for 4 h. Finally, the boron nitride nanofibers (BNNF) were obtained and preserved for the preparation of nanocomposites.

2.3. Preparation of Epoxy-based Nanocomposites Film

According to a proportion of epoxy/hardener/accelerant of 1/0.86/0.02, the weight of BNNF we needed was calculated (0.5 wt%, 1 wt% and 2 wt%). First, epoxy was poured into a beaker and added to the BNNF, and was subjected to high-speed mechanical stirring until the nanofibers were dispersed well; then the hardener and the accelerant were poured in and stirred at a high speed. The mixture was put into the vacuum oven for removing bubbles at 60 °C for 20 min.

Before the automatic coating machine was used, the machine should lay a layer of polyethylene film and be preheated at 100 °C—the thickness of the coating was 100 μm. Next, the air pump was turned on, which made the polyethylene film attach tightly to the machine. The moderate BNNF/epoxy suspension was wiped by the automatic coating machine and poured into a hollow model with a thickness of 100 micron, a polyethylene film was attached to the suspension to prevent the epoxy nanocomposites shrinking. The film was heated at 100 °C until the suspension was completely dried. After that, the film was transferred into the oven for curing. The procuring was held at 100 °C for 2 h; the post curing was held at 150 °C for 10 h. Finally, when the temperature cooled down to room temperature, the EP/BNNF film was taken out with the polyethylene film and the EP/BNNF film was carefully stripped out.

2.4. Characterization

The morphological structures of the fillers and the surface morphology and fractured morphology of the composites were characterized using field emission scanning electron microscopy (FESEM, FEI QUANTA F250, Hillsboro, TX, USA). The microstructure of the BNNF and EP/BNNF film was observed with an FEI Tecnai G2 F20 S-TWIN (Hillsboro, TX, USA) transmission electron microscope (TEM). The phase structure of the prepared BNNF was tested with X-ray diffraction (XRD) using a Bruker D2 PHASER diffractometer (Karlsruhe, Germany). The thermal diffusivity (α) and specific heat (C_p) were measured with an LFA 467 Nanoflash (NETZSCH, Selb, Germany). The bulk density (ρ) of the sample was given by the product of the length, width and height. The thermal conductivity value of the samples was calculated by the product of the thermal diffusivity, specific heat, and bulk density.

The equation is $\lambda = \alpha \rho C_p$. In the test of characterization, the samples were cut into a quadrate shape, 10×10 mm^2 in proportion and 0.3 mm in thickness. For the dielectric property measurements, gold electrodes were first sputtered on both sides of the epoxy-based composites by the auto fine coater with a mask that had 3 mm diameter eyelets. The dielectric properties of the composites were measured with a broadband dielectric spectrometer (CONCEPE 80, Novocontrol Technology Company, Montabaur, Germany) with an Alpha-A high-performance frequency analyzer over the frequency range of 10^{-1} Hz to 10^6 Hz. Thermal gravimetric analysis (TGA) was performed with a heating rate of 60 °C/min under nitrogen. Differential scanning calorimetry (DSC, 200 F3, NETZSCH, Selb, Germany) was performed at temperatures from 25 °C to 200 °C at a heating rate of 10 °C/min under a nitrogen atmosphere to study the glass transition temperature (T_g) of the composites. The electrical breakdown strength at the commercial power frequency of composites was measured using the standard of IEC 60243.

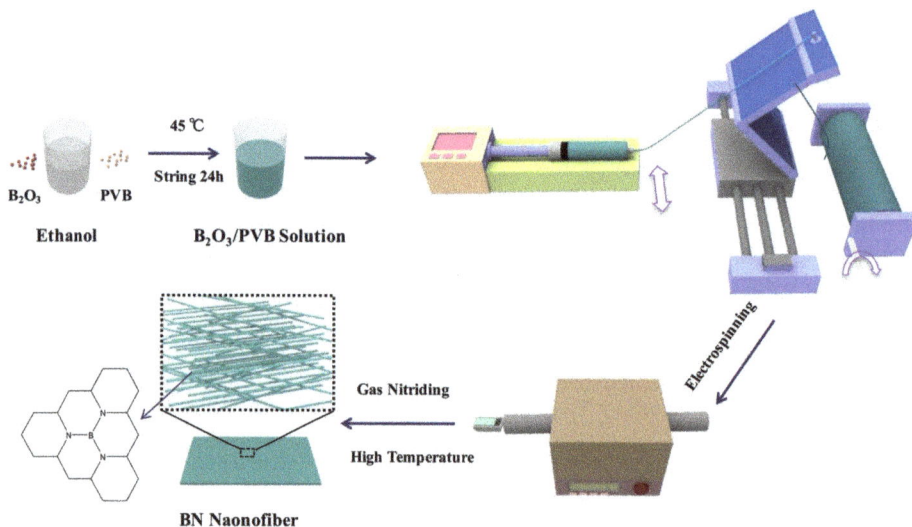

Figure 1. Schematic illustration showing the fabrication of boron nitride (BN) fiber via electrospinning.

3. Results and Discussion

3.1. X-ray Diffraction, SEM and TEM Analysis (BNNF)

We utilized the scanning electron microscope (SEM), transmission electron microscope (TEM) and X-ray diffraction (XRD) to characterize the morphology and elements of PVB/B$_2$O$_3$ composite fiber and the prepared boron nitride nanofiber. As shown in Figure 2a, PVB/B$_2$O$_3$ composite fiber is thin and has a smooth morphology with a uniform diameter ranging from 150 nm to 200 nm, when the flow rate is 1 mL/h. Figure S1a–f shows the morphology of PVB/B$_2$O$_3$ fiber prepared with a different flow rate. According to these SEM photos, it was found that in the circumstance of the same solution proportion and voltage, when the flow rate is more than 1 mL/h, the surface of the fiber becomes coarser and forms a structure of beads and the diameter of the fiber tends to be bigger with increasing flow rate. Figure 2b shows the SEM photograph of BNNF (inset is an enlarged view). It can be seen that the surface of the BN fiber is smooth and the average diameter of a BN fiber is around 150 nm (see Figure S2). Figure 2c shows TEM images of BN nanofibers with an high resolution transmission electron microscope (HRTEM) image in the inset. It can be seen that the lattice spacing is 0.33 mm in its high-resolution image, which is the typical value of BN lattice spacing [40]. The XRD pattern of BNNF is shown in Figure 2d. BN fibers exhibited a well-crystallized hexagonal structure. The peaks at approximately 26.9, 41.6, and 54.9 are assigned to the (002), (100), (004) and (110) crystallographic planes of h-BN, respectively [41]. These peaks are typical of h-BN according to the MDI Jade database. There were no other impurity phases being detected, also indicating the high purity of BN nanofibers. The high crystalline quality of BN fillers was beneficial to improving the thermal conductive and electrical properties of the composites.

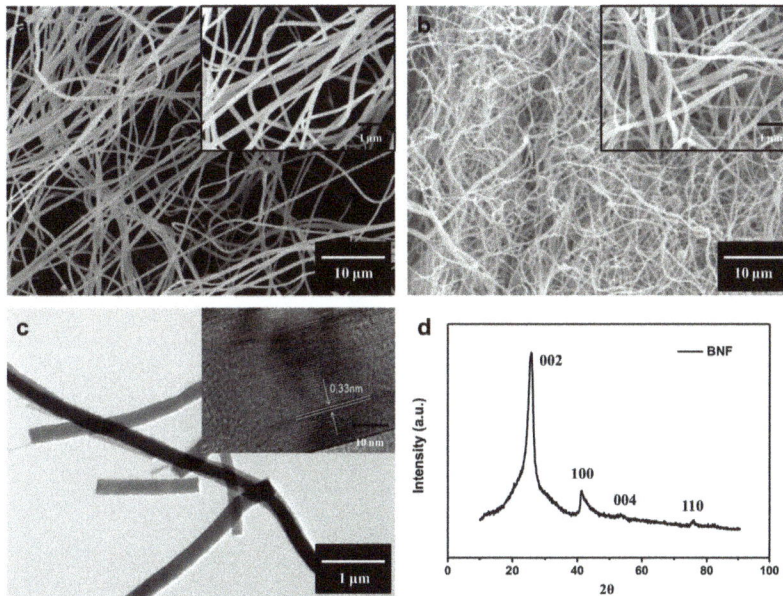

Figure 2. Scanning electron microscopy (SEM) image of (**a**) B_2O_3/PVB composite fiber and (**b**) BN fiber samples with insets showing the corresponding enlarged views. (**c**) Transmission electron microscopy (TEM) images of BN fiber with HRTEM image in the inset. (**d**) X-ray diffraction (XRD) pattern of BN fiber.

3.2. SEM and Optical Microscope Analysis of BNNF/Epoxy Composites

Figure 3a,b show the cross images of BNNF/EP film with 1 wt% BNNF and 2 wt% BNNF, respectively. According to Figure 3a, the thread ends of BNNF can be seen and there are some hollows (black arrows) dispersed on the cross section, which were caused by fractured BN nanofibers in liquid nitrogen, while BNNF nanofibers were uniformly dispersed in the epoxy matrix. Recently, the alignment of 1D and 2D nanomaterials has been widely applied to prepare anisotropic functional materials. For instance, boron nitride nanosheets and graphene polymer composites can dramatically increase thermal conductivity in the in-plane direction because of their anisotropic performance. The A.A. Balandin research group aligned the graphene functionalized with Fe_3O_4 nanoparticles by applying an external magnetic field. The thermal conductivity of composites with low loading oriented graphene reaches up to 1.3 W/m·k [42]. G. Lazzara et al. developed Halloysite as a tubular template to obtain ordered arrays of clay nanotubes on solid substrates with multifunction including controlled release, good mechanical properties and anti-water [43–46]. Indeed, many methods such as filtration, freezing ice, hot pressing, shear or extrusion flow, stretching, electrospinning, electric field, magnetic field and so forth, have used to acquire the oriented nanofillers in polymer matrix. In this work, the casting method is applied to align the BNNF in the epoxy matrix via drawknife. The drawknife provided the propelling force on the side of nanofibers, and compelling the fiber to lie down in the epoxy matrix. In the other words, the lying fibers in the polymer were more stable than the standing fibers when they were loaded the planus force, perpendicular to the standing direction. Most thread ends of BNNF in epoxy are perpendicular to the cross section of the composite, which demonstrated the good orientation of BNNF. The reason is that the drawknife provided an external force on the BN fiber in composites during the coating of the suspension with epoxy and BNNF, resulting in the formation of a BNNF carpet [34]. In other words, the drawknife produced a push force on BNNF, leading to the fall of BNNF perpendicular to the drawknife. Moreover, the clear fiber-like structure (BN fiber) was observed by optical microscope and their length was within the region of 4–9 μm. Optical microscope images of BNNF/epoxy composites (Figure 3c) are more macroscopic to show the uniform dispersion and alignment of BNNF (white arrows showing the aligned orientation) in the epoxy matrix. To exhibit the good flexibility and transparency properties of our BNNF-epoxy film, we folded the film (inset in Figure 3a) with tweezers over the ruler, which demonstrated the good mechanical properties and homogeneous dispersion of BNNF in composites. As shown in Figure 3b, some BN nanofibers formed agglomerations in composites due to the high content and surface area and energy of BNNF. Correspondingly, the partial agglomerations of BNNF in epoxy composites with 2 wt% BNNF can also be found in the optical microscope images of the BNNF/epoxy composites (Figure 3d). Despite all this, we can still find the obvious alignment of BNNF (as indicated by the white arrows) in epoxy composites.

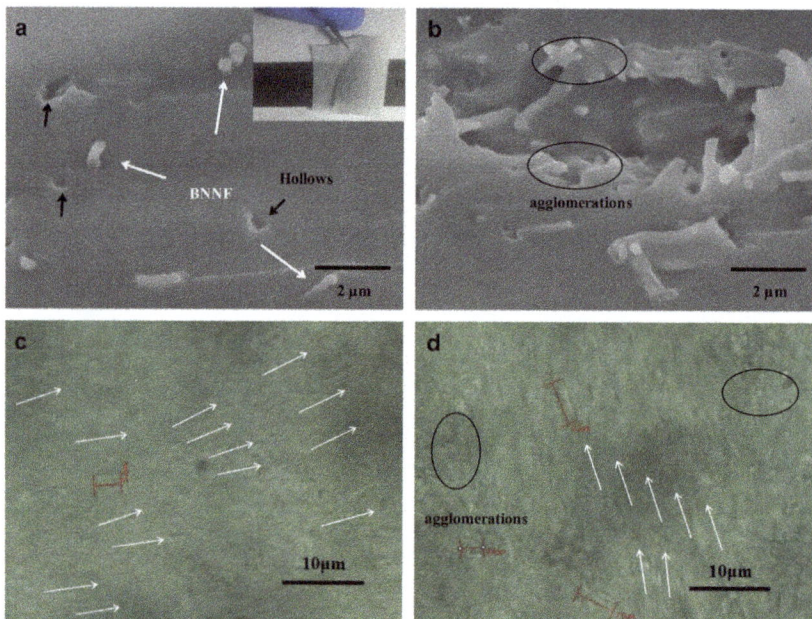

Figure 3. SEM cross-section image of (**a**) 1 wt% BNNF/EP and (**b**) 2 wt% BNNF/EP film. The inset in (**a**) shows the flexibility and transparency of sample. Optical microscope surface photograph of (**c**) 1 wt% BNNF/EP and (**d**) 2 wt% BNNF/EP composite film.

3.3. Dielectric and Thermal Properties

As expected, the dielectric permittivity increased with the increment of the filler loading in epoxy over all frequency ranges from 10^{-1}–10^6 Hz at room temperature (298 K) for the epoxy composites with different BNNF mass fractions. Figure 4a,b show the relationship between permittivity(ε') and dielectric loss(tanδ) of BNNF/epoxy composites film with various BNNF mass fraction and frequency. In BNNF/epoxy composite systems, mainly 3 parts of dielectric loss are included as follows: (1) the conduction loss caused by incorporated inorganic fillers; (2) the relaxation loss gave rise to the polar functional groups of polymer matrix; (3) interface loss from the interface between filler and polymer matrix. According to Figure 4a, as the frequency increased, the permittivity of all the samples tended to decrease. It can be attributed to the lower influence of relaxational polarization, which was unable to keep pace with the high frequency change so that displacement polarization became dominant. Therefore, the permittivity of composites decreased. The permittivity of pure epoxy film was around 3.5 under the frequency of 1 kHz and the permittivity of most BNNF/EP composite samples were slightly higher. This would be explained by the interfaces among fillers and epoxy resin interacting to cause interfacial polarization, which is a significant factor for the increase of permittivity of composite material. As shown in Figure 4a, in low-frequency regions, the value of permittivity of composite increased as increasing the amount of BNNF. This could be interpreted as the larger the content of BNNF, the comparatively larger the interface boundary between filler and matrix; thus, the interfacial effect and polarization became stronger so that the permittivity became slightly higher. However, in high-frequency regions, the relaxation polarization in the composite could not follow the change of frequency [7]. As the influence of interface polarization on the permittivity was less, the value of permittivity of all the samples decreased and tended to continually decrease with increasing frequency. What is more, the dielectric constant of the composite with 1 wt% BNNF is lower than that of other composites in high frequency zones. The high-frequency dielectric response can be accounted for by the C-F dipole orientation polarization of the composites based on the Debye relaxation theory [13]. The dielectric properties of polymer incorporated by semiconductors or insulators have close contact with the electrical properties of inorganic fillers, significantly affecting the accumulation and migration of free carriers at the interaction area between the filler and the polymer matrix. The dielectric constant of BNNF/epoxy composites is lower than that of pure epoxy.

Figure 4b presents the dependence of electric conductivity on frequency at room temperature of BNNF/epoxy composites. The electric conductivity of all samples increases with the increment of the tested frequency. The dielectric constant of the composites with BNNF is lower than that of neat epoxy at all tested frequency ranges from 10^{-1} to 10^6 Hz. The reasons can be ascribed to the high intrinsic resistance of BNNF and the interface layer between polymer and fillers according to the core-shell theory [9], which blocks the direct transfer of free charge carriers and prolongs the transport path of carriers in the epoxy matrix. Therefore, one can declare that the incorporating BNNF plays a crucial role in suppressing the conduction of the epoxy composites.

The breakdown field strength (BD) and its Weibull distribution of the breakdown test of epoxy nanocomposites with 0.5 wt%, 1 wt%, 2 wt% BNNF and pure epoxy are shown in Figure 4c. According to the experimental results in Figure 4c, the BD strength of the epoxy composites with 0.5 wt% BNNF was slightly higher than that of pure epoxy film; the BD strength of the sample of 1 wt% BNNF was higher obviously; but the BD strength of the sample of 2 wt% BNNF was slightly lower than that of the pure epoxy film. In general, that doping micro-sized inorganic fillers into polymer would decrease the BD strength of the polymer itself [2,11]. First, surface energy of micro-sized filler is weak and polymer molecule cannot closely attach to micro filler, leading to the obvious incompatible interface between filler and polymer matrix; Second, dielectric permittivity of inorganic filler is much higher than that of polymer which would cause large-scale electric field distortion in the polymer composite, leading to the decrease of BD strength; The other reason is that the voids and impurity caused by the doping fillers make the BD strength of composite lower than that of pure polymer. However, in this work, specific surface area of BNNF is large. Actually, the large surface area of nanomaterials has many applications such as battery, wave absorption and electromagnetic properties and so forth [47–56]. In other words, BNNF has high surface energy, which is beneficial to form closer interface. In addition, it is worth noting that BNNF has lower dielectric constant in comparison with other high thermal conductive inorganic filler (such as AlN, Al_2O_3 and SiC). Dielectric constant of BNNF approaches to that of pure epoxy, and the close permittivity between BNNF and epoxy relieves field distortion of BNNF/epoxy composites. Moreover, BNNF prepared by our method is very purified and has no other impurity phases, preventing the effect on BD from impurity phases. In addition, the number and size of voids and bubbles are effectively reduced by using hot-pressing technology. From our results, the moderate amount of doping BNNF in epoxy resin can increase the BD strength of composite. It can be explained that the doping BNNF is directionally distributed near the interface of the film and the oriented nanofibers of tubule are perpendicular to the electric field direction so that will signally scatter electrons, which could effectively restrain the growth of electrical tree and extend the growth path of electrical tree thus put off the forming of penetrating breakdown. Therefore, the BD strength of the composite increases. Nevertheless, as the mass fraction of BNNF continually increases, the BD strength of composites decreases. It is attributed to the more amount of fillers, the more aggregation of nanofibers in the composites and the more difficulty of fillers distribution in matrix, resulting in decrease in the BD strength of the samples.

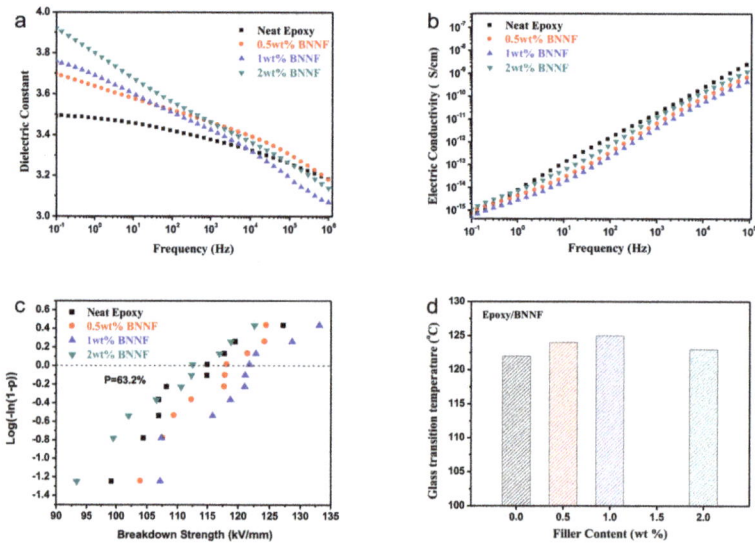

Figure 4. (**a**) Dielectric permittivity(ε') and (**b**) electric conductivity dependence on frequency and filler loading for the epoxy composite film with different mass fraction of BNNF at room temperature. (**c**) Weibull plots of breakdown strength of Epoxy/BNNF composite film. (**d**) DSC curves of four samples of different mass fraction of BNNF.

The DSC curves of pure epoxy, 0.5 wt%, 1 wt% and 2 wt% BNNF film are shown in Figure 4d. The glass transition temperature (T_g) of the pure epoxy film is around 121 °C Compared with that of epoxy, the T_g of epoxy composites with 0.5 wt% and 1wt% BNNF is 125 °C and 126 °C, respectively (See Figure S2). However, the Tg of epoxy composite with 2 wt% BNNF shows slightly decreased in comparison with that of composites with lower BNNF mass fraction. It can be attributed to the uniform dispersion of BNNF in epoxy matrix (see Figure 3a,c), resulting the more interface between filler and epoxy. In addition, a reason for this observation is attributed to the interfacial interaction between BN fiber fillers embedded in composites and epoxy matrix, which restricts the movement of the polymer molecular chain and restrains the thermal decomposition, leading to the higher T_g and increasing the thermal stability of epoxy matrix (see Figure S3). The BNNF has decent thermal conductivity and thermal stability properties, which can effectively transfer heat. Meanwhile, the measuring C_p data and their accuracy are summarized in Table S1 (See the Supporting information). The Cp of pure epoxy is higher than that of BNNF/epoxy composites. It is attributed to the good mobility of neat polymer. BNNF incorporated into polymer retained the free mobility of the polymer, resulting in lower C_p. Moreover, the intrinsic C_p of BNNF is lower in comparison with that of epoxy matrice.

3.4. Thermal Conductivity

The as-fabricated BNNF/epoxy nanocomposite films were evaluated for their thermal transport properties by using a commercially acquired instrument based on the Laser flash method [37,38], measuring the thermal diffusivities (TD) and specific heat (C_p) in each film. It is worth noting that thin gold electrodes (around 10 nm) need to be sputtered on both sides of films by the auto coater to delay laser transmitting time. In addition, a spot of graphite was sprinkled on the surface of gold coating for receiving the heat from laser. The accuracy of the instrument for specimen of TD and C_p was evaluated and validated by measuring and comparing the standard glass (Pyrex7740), from which the average deviations with respect to the use of different sets of instrumental parameters and in repeated measurements were generally less than 10%. For the BNNF/epoxy nanocomposite films, multiple measurements of several specimens yielded TD values shown in Table 1 and corresponding C_p and density data were shown in Table S1. When the temperature is constant at T, the calculation relationship between thermal conductivity and TD is as follows [5]:

$$\lambda(T) = \alpha(T) \cdot C_p(T) \cdot \rho(T) \tag{1}$$

where λ is thermal conductivity, α is thermal diffusivity, C_p is specific heat and ρ is bulk density of sample.

As expected, the thermal diffusivities of the polymeric nanocomposites are strongly dependent on the BNNF content, with higher mass fractions consistently resulting in higher thermal diffusivities of the nanocomposites. Figure 5 shows the dependence of the thermal diffusivity on the mass fraction of nanofiber in the BNNF/epoxy nanocomposite films. From the Figure 5, as the mass fraction of BNNF increased, the thermal diffusivity of the composites increased obviously. Meanwhile, thermal conductivity of nanocomposites increased with increment of BNNF. The thermal conductivity value of composite with 2 wt% BNNF increased to 0.205 W/mK (shown in Table 1 and their accuracy shown in Table S1), which enhanced 26% than that of pure epoxy. It is well known that the heat conduction mechanism of BN is phonon transfer which caused by lattice vibration. As the mass fraction of filler increased, the quantity of microcosmic interface between filler and epoxy-base increased, which increased the chance of phonon scattering that restrained the transfer of phonon. Nevertheless, when the fillers were added to a certain amount that long nanofibers were easier to connect to each other, it would form a partial thermal transport network.

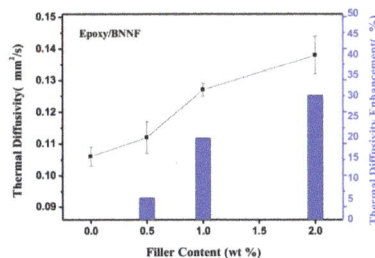

Figure 5. The thermal diffusivity and enhancement of four samples with different mass fraction of BNNF.

Table 1. Thermal diffusivity and dielectric breakdown strength of samples with different mass fraction of BNNF.

Samples	λ (W/m·k)	Nielsen (W/m·k)	α (cm²/s)	DBS (kV/mm)	β	Thickness (μm)	Deviation (μm)
Pure Epoxy	0.162	0.162	0.106	115.8	14.74	112	±8
0.5 wt%	0.17	0.181	0.112	118.8	20.95	108	±3
1 wt%	0.185	0.194	0.127	122.8	19.67	112	±7
2 wt%	0.205	0.227	0.138	113.5	15.05	115	±3

λ: thermal conductivity α: thermal diffusion coefficient, DBS: Dielectric Breakdown Strength, β: Shape parameter.

Nielsen proposed a relatively simple model for the evaluation of thermal conductivities of a composite material [57]. In his approach, the thermal conductivity of a composite material, Kc, is related to the thermal conductivity of a matrix, K_m, and a filler, K_f, according to the following equation:

$$K_c = K_m[(1 + ABV_f)/|1 - B\beta V_f|] \tag{2}$$

where the parameters B and β are given by

$$B = [(K_f/K_m - 1)/(K_f/K_m + A)] \tag{3}$$

$$\beta = [(1 - V_m)/V_m^2]V_f + 1 \tag{4}$$

here, V_f represents the volume fraction of the filler, A is a constant related to the generalized Einstein coefficient reported for most of the materials, [57] and V_m is the maximum packing fraction.

The experimental thermal conductivity values for BNNF/epoxy composites are in agreement with the calculated results by the Nielsen's model (see Table 1). It is worth noting that Terao et al. have reported on the BNNTs/PVA composites [58]. In that report, The BN nanotubes were used as the fillers and the thermal conductivity behavior of composites incorporating oriented and random fillers was researched by using the Nielsen's model. In their research, the theoretical results were significant different with experimental results. The theoretical results showed that thermal conductivity sharply increased when the BNNTs content in a composite with randomly dispersed BNNTs was more than 30 vol%. However, the similar increase took place when the BNNTs fraction exceeded 10 vol% (for the aligned BNNT case). In other words, the percolating value for the increase in thermal conductivity decreased to 10 vol% when BNNTs were oriented in composites. Therefore, alignment of one-dimensional fillers such as BNNTs and CNTs can enhance thermal conductivity of composites under the condition of containing same fillers volume fraction [58,59]. It can be attributed to closer distance between fillers and the decrease of percolating fraction in polymer. Moreover, the low volume fraction of fillers incorporated in polymers can enhance the mechanical and electrical properties of polymers in comparison with the high content filler in composites. The theory indicates that the thermal conductivities of BNNF/epoxy composites have still big room for improvement of if one manages to incorporate higher fractions of BNNF into a polymer while maintaining good alignment of the fillers.

4. Conclusions

In summary, the addition of BNNF to the epoxy composites has no obvious effect on the dielectric permittivity of epoxy. However, the electric conductivity of the composites was lower than that of pure epoxy within the tested frequency region due to the high intrinsic resistance of BNNF and the interface layer between polymer and fillers, which blocks the direct transfer of free charge carriers and prolong the transport path of carriers in epoxy matrix. Adding moderate mass fraction of BNNF improved the breakdown strength of the composites: the breakdown strength of epoxy composite film with 1 wt% BNNF was 122.9 kV/mm, increased 4.8% compared with that of pure epoxy film (117.2 kV/mm);

the addition of BNNF to the composites also influenced its thermal properties. The interfacial effect was formed after doping the fillers into the epoxy resin, which improved the heat transmission temperature and the thermal stability of the composites. Furthermore, due to the superior heat-conducting property of BN itself, the thermal conductivity values of the BNNF/epoxy composite films were enhanced. The thermal conductivity of 2 wt% BNNF/epoxy film was increased 26.4% compared with the pure epoxy film.

Supplementary Materials: The following are available online at http://www.mdpi.com/2079-4991/8/4/242/s1, Figure S1: The Proportion of PVB/B_2O_3 is 700 mg/350 mg, the voltage is 16.14 kV, the flow is (a) 0.75 mL/h (b) 1.5 mL/h (c) 2.0 mL/h (d) 2.5 mL/h (e) 4.0 mL/h (f) 1.0 mL/h. Figure S2: DSC curves of the BNNF/epoxy composites with different filler content under nitrogen atmosphere. Figure S3: TGA curves of the BNNF/epoxy composites with different filler content under nitrogen atmosphere. Table S1: A summary of the detailed data information on thermal conductivity.

Acknowledgments: The authors thank the colleagues in the laboratory of international Center for Dielectric Research for their support. Financial support was provided by the National Natural Science Foundation of China (No. 51407134), (Innovative Research Group, No. 51521065), China Postdoctoral Science Foundation (No. 2016M590619), Natural Science Foundation of Shandong Province (No. ZR2016EEQ28), Qingdao Postdoctoral Application Research Prohect and State Key Laboratory of Electrical Insulation and Power Equipment (No. EIPE14107).

Author Contributions: Zhengdong Wang and Jingya Liu conceived, designed and proformed the experiments; Zhengdong Wang, Yonghong Cheng, Siyu Chen, Mengmeng Yang, Jialiang Huang and Hongkang Wang analyzed the data and supervised research; Guanglei Wu and Hongjing Wu also gave some good advices during the preparation of BNNF via electrospinning. Zhengdong Wang and Jingya Liu wrote the paper together.

Conflicts of Interest: The authors declare no conflict of interest.

References

1. Huang, X.; Zhi, C.; Jiang, P.; Golberg, D.; Bando, Y.; Tanaka, T. Polyhedral oligosilsesquioxane-modified boron nitride nanotube based epoxy nanocomposites: An ideal dielectric material with high thermal conductivity. *Adv. Funct. Mater.* **2013**, *23*, 1824–1831. [CrossRef]

2. Wang, Z.; Cheng, Y.; Wang, H.; Yang, M.; Shao, Y.; Chen, X.; Tanaka, T. Sandwiched epoxy-alumina composites with synergistically enhanced thermal conductivity and breakdown strength. *J. Mater. Sci.* **2017**, *52*, 4299–4308. [CrossRef]

3. Wu, G.; Cheng, Y.; Yang, Z.; Jia, Z.; Wu, H.; Yang, L.; Li, H.; Guo, P.; Lv, H. Design of Carbon Sphere/Magnetic Quantum Dots with Tunable Phase Compositions and Boost Dielectric Loss Behavior. *Chem. Eng. J.* **2018**, *333*, 519–528. [CrossRef]

4. Hou, Z.-L.; Song, W.-L.; Wang, P.; Meziani, M.J.; Kong, C.Y.; Anderson, A.; Maimaiti, H.; LeCroy, G.E.; Qian, H.; Sun, Y.-P. Flexible Graphene-Graphene Composites of Superior Thermal and Electrical Transport Properties. *ACS Appl. Mater. Interf.* **2014**, *6*, 15026–15032. [CrossRef] [PubMed]

5. Veca, L.M.; Meziani, M.J.; Wang, W.; Wang, X.; Lu, F.; Zhang, P.; Lin, Y.; Fee, R.; Connell, J.W.; Sun, Y.-P. Carbon Nanosheets for Highly Thermal Conductive Polymeric Nanocomposites. *Adv. Mater.* **2009**, *21*, 2088–2092. [CrossRef]

6. Huang, Y.; Min, D.; Li, S.; Wang, X.; Lin, S. Dielectric relaxation and carrier transport in epoxy resin and its microcomposite. *IEEE Trans. Dielectr. Electr. Insul.* **2017**, *24*, 3083–3091. [CrossRef]

7. Huang, Y.; Min, D.; Li, S.; Li, Z.; Xie, D.; Wang, X.; Lin, S. Surface flashover performance of epoxy resin microcomposites improved by electron beam irradiation. *Appl. Surf. Sci.* **2017**, *406*, 39–45. [CrossRef]

8. Song, W.-L.; Wang, P.; Cao, L.; Anderson, A.; Meziani, M.J.; Farr, A.J.; Sun, Y.-P. Polymer/Boron Nitride Nanocomposite Materials for Superior Thermal Transport Performance. *Angew. Chem. Int. Ed.* **2012**, *51*, 6498–6501. [CrossRef] [PubMed]

9. Wang, Z.; Cheng, Y.; Yang, M.; Huang, J.; Cao, D.; Chen, S.; Xie, Q.; Lou, F.; Wu, H. Dielectric properties and thermal conductivity of epoxy composites using core/shell structured Si/SiO_2/Polydopamine. *Compos. Part B Eng.* **2018**, *140*, 83–90. [CrossRef]

10. Wang, Z.; Cheng, Y.; Shao, Y.; Xie, Q.; Wu, G. Thermal conductivity and electric breakdown strength properties of epoxy/alumina/boron nitride nanosheets composites. In Proceedings of the 2016 IEEE International Conference on Dielectrics (ICD), Montpellier, France, 3–7 July 2016; pp. 355–358. [CrossRef]

11. Wang, Z.B.; Iizuka, T.; Kozako, M.; Ohki, Y.; Tanaka, T. Development of epoxy/BN composites with high thermal conductivity and sufficient dielectric breakdown strength Part I. Sample preparations and thermal conductivity. *IEEE Trans. Dielectr. Electr. Insul.* **2011**, *18*, 1963–1972. [CrossRef]

12. Zhou, Y.C.; Yao, Y.; Chen, C.Y.; Moon, K.; Wang, H.; Wong, C.P. The use of polyimide-modified aluminum nitride fillers in AlN@PI/Epoxy composites with enhanced thermal conductivity for electronic encapsulation. *Sci. Rep.* **2014**, *4*, 4779. [CrossRef] [PubMed]

13. Gu, J.; Lv, Z.; Wu, Y.; Guo, Y.; Tian, L.; Qiu, H.; Li, W.; Zhang, Q. Dielectric thermally conductive boron nitride/polyimide composites with outstanding thermal stabilities via in situ polymerization-electrospinning-hot press method. *Compos. Part A Appl. Sci. Manuf.* **2017**, *94*, 209–216. [CrossRef]

14. Pan, C.; Zhang, J.; Kou, K.; Zhang, Y.; Wu, G. Investigation of the through-plane thermal conductivity of polymer composites with in-plane oriented hexagonal boron nitride. *Int. J. Heat Mass Transf.* **2018**, *120*, 1–8. [CrossRef]

15. Meziani, M.J.; Song, W.-L.; Wang, P.; Lu, F.; Hou, Z.; Anderson, A.; Maimaiti, H.; Sun, Y.-P. Boron Nitride Nanomaterials for Thermal Management Applications. *ChemPhysChem* **2015**, *16*, 1339–1346. [CrossRef] [PubMed]

16. Chen, J.; Huang, X.; Zhu, Y.; Jiang, P. Cellulose Nanofiber Supported 3D Interconnected BN Nanosheets for Epoxy Nanocomposites with Ultrahigh Thermal Management Capability. *Adv. Funct. Mater.* **2017**, *27*, 1604754. [CrossRef]

17. Nika, D.L.; Balandin, A.A. Phonons and thermal transport in graphene and graphene-based materials. *Rep. Prog. Phys. Phys. Soc.* **2017**, *80*, 036502. [CrossRef] [PubMed]

18. Saadah, M.; Hernandez, E.; Balandin, A. Thermal Management of Concentrated Multi-Junction Solar Cells with Graphene-Enhanced Thermal Interface Materials. *Appl. Sci.* **2017**, *7*, 589. [CrossRef]

19. Shahil, K.M.F.; Balandin, A.A. Thermal properties of graphene and multilayer graphene: Applications in thermal interface materials. *Solid State Commun.* **2012**, *152*, 1331–1340. [CrossRef]

20. Li, Z.; Li, Z.A.; Sun, S.; Zheng, D.; Wang, H.; Tian, H.; Yang, H.; Bai, X.; Li, J. Direct Observation of Inner-Layer Inward Contractions of Multiwalled Boron Nitride Nanotubes upon in Situ Heating. *Nanomaterials* **2018**, *8*, 86. [CrossRef] [PubMed]

21. Liu, J.-H.; Yang, S.-T.; Wang, X.; Wang, H.; Liu, Y.; Luo, P.G.; Liu, Y.; Sun, Y.-P. Carbon Nanoparticles Trapped in vivo—Similar to Carbon Nanotubes in Time-Dependent Biodistribution. *ACS Appl. Mater. Interfaces* **2014**, *6*, 14672–14678. [CrossRef] [PubMed]

22. Wang, Y.; Mortimer, M.; Chang, C.H.; Holden, P. Alginic Acid-Aided Dispersion of Carbon Nanotubes, Graphene, and Boron Nitride Nanomaterials for Microbial Toxicity Testing. *Nanomaterials* **2018**, *8*, 76. [CrossRef] [PubMed]

23. Zhi, C.; Bando, Y.; Terao, T.; Tang, C.; Kuwahara, H.; Golberg, D. Boron Nanotube–Polymer Composites: Towards Thermoconductive, Electrically Insulating Polymeric Composites with Boron Nitride Nanotubes as Fillers. *Adv. Funct. Mater.* **2009**, *19*, 1857–1862. [CrossRef]

24. Chopra, N.G.; Luyken, R.J.; Cherrey, K.; Crespi, V.H.; Cohen, M.L.; Louie, S.G.; Zettl, A. Boron nitride nanotubes. *Science* **1995**, *269*, 966. [CrossRef] [PubMed]

25. Ahmad, P.; Khandaker, M.U.; Amin, Y.M. Synthesis of boron nitride nanotubes by Argon supported Thermal Chemical Vapor Deposition. *Phys. E Low-Dimens. Syst. Nanostruct.* **2015**, *67*, 33–37. [CrossRef]

26. Kalay, S.; Yilmaz, Z.; Sen, O.; Emanet, M.; Kazanc, E.; Çulha, M. Synthesis of boron nitride nanotubes and their applications. *Beilstein J. Nanotechnol.* **2015**, *6*, 84–102. [CrossRef] [PubMed]

27. Huang, X.; Jiang, P. Core–Shell Structured High-k Polymer Nanocomposites for Energy Storage and Dielectric Applications. *Adv. Mater.* **2015**, *27*, 546. [CrossRef] [PubMed]

28. Wang, Z.; Qu, S.; Cheng, Y.; Zheng, C.; Chen, S.; Wu, H. Facile synthesis of Co_3O_4, spheres and their unexpected high specific discharge capacity for Lithium-ion batteries. *Appl. Surf. Sci.* **2017**, *416*, 338–343. [CrossRef]

29. Wang, G.; Yu, D.; Kelkar, A.D.; Zhang, L. Electrospun Nanofiber: Emerging Reinforcing Filler in Polymer Matrix Composite Materials. *Prog. Polym. Sci.* **2017**, *75*. [CrossRef]

30. Wang, J.; Wang, H.; Cao, D.; Lu, X.; Han, X.; Niu, C. Epitaxial Growth of Urchin-Like CoSe$_2$ Nanorods from Electrospun Co-Embedded Porous Carbon Nanofibers and Their Superior Lithium Storage Properties. *Part. Part. Syst. Charact.* **2017**, *34*, 1700185. [CrossRef]

31. Ko, F.; Gogotsi, Y.; Ali, A.; Naguib, N.; Ye, H.; Yang, G.L.; Li, C.; Willis, P. Electrospinning of Continuous Carbon Nanotube-Filled Nanofiber Yarns. *Adv. Mater.* **2003**, *15*, 1161–1165. [CrossRef]

32. Ren, L.; Pashayi, K.; Fard, H.R.; Hotha, P.S.; Tasciuc, T.B.; Ozisik, T.R. Engineering the coefficient of thermal expansion and thermal conductivity of polymers filled with high aspect ratio silica nanofibers. *Compos. Part B Eng.* **2014**, *66*, 228–234. [CrossRef]

33. Jae-Kon, L.; Jin-Gon, K. Generalized Self-Consistent Model for Predicting Thermal Conductivity of Composites with Aligned Short Fibers. *Mater. Trans.* **2010**, *51*, 2039–2044. [CrossRef]

34. Hu, P.; Wang, J.; Shen, Y.; Guan, Y.; Lin, Y.; Nan, C. Highly enhanced energy density induced by hetero-interface in sandwich-structured polymer nanocomposites. *J. Mater. Chem. A* **2013**, *1*, 12321–12326. [CrossRef]

35. Rana, D.; Sauvant, V.; Halary, J.L. Molecular analysis of yielding in pure and antiplasticized epoxy-amine thermosets. *J. Mater. Sci.* **2002**, *37*, 5267–5274. [CrossRef]

36. Feng, A.; Jia, Z.; Yu, Q.; Zhang, H.; Wu, G. Preparation and Characterization of Carbon Nanotubes/Carbon Fiber/Phenolic Composites on Mechanical and Thermal Conductivity Properties. *NANO* **2018**. [CrossRef]

37. Rana, D.; Mounach, H.; Halary, J.L.; Monnerie, L. Differences in mechanical behavior between alternating and random styrene-methyl methacrylate copolymers. *J. Mater. Sci.* **2005**, *40*, 943–953. [CrossRef]

38. Feng, A.; Wu, G.; Pan, C.; Wang, Y. Synthesis, preparation and mechanical property of wood fiber-reinforced poly(vinyl chloride) composites. *J. Nanosci. Nanotechnol.* **2017**, *17*, 3859–3863. [CrossRef]

39. Feng, A.; Wu, G.; Pan, C.; Wang, Y. The Behavior of Acid Treating Carbon Fiber and the Mechanical Proper ties and Thermal Conductivity of Phenolic Resin Matrix Composites. *J. Nanosci. Nanotechnol.* **2017**, *17*, 3786–3791. [CrossRef]

40. Qiu, Y.; Yu, J.; Rafique, J.; Yin, J.; Bai, X.; Wang, E. Large-Scale Production of Aligned Long Boron Nitride Nanofibers by Multijet/Multicollector Electrospinning. *J. Phys. Chem. C* **2009**, *113*, 11228–11234. [CrossRef]

41. Coleburn, N.L.; Forbes, J.W. Irreversible Transformation of Hexagonal Boron Nitride by Shock Compression. *J. Chem. Phys.* **1968**, *48*, 555–559. [CrossRef]

42. Renteria, J.; Legedza, S.; Salgado, R.; Balandin, M.P.; Ramirez, S.; Saadah, M.; Kargar, F.; Balandin, A.A. Magnetically-functionalized self-aligning graphene fillers for high-efficiency thermal management applications. *Mater. Des.* **2015**, *88*, 214–221. [CrossRef]

43. Lazzara, G.; Cavallaro, G.; Panchal, A.; Fakhrullin, R.; Stavitskaya, A.; Vinokurov, V.; Lvov, Y. An assembly of organic-inorganic composites using halloysite clay nanotubes. *Curr. Opin. Colloid Interface Sci.* **2018**, *35*. [CrossRef]

44. Lee, J.W.; Park, S.J.; Kim, Y.H. Improvement of Interfacial Adhesion of Incorporated Halloysite-Nanotubes in Fiber-Reinforced Epoxy-Based Composites. *Appl. Sci.* **2017**, *7*, 441. [CrossRef]

45. Cavallaro, G.; Danilushkina, A.A.; Evtugyn, V.G.; Lazzara, G.; Milioto, S.; Parisi, F.; Parisi, E.V.; Fakhrullin, R.F. Halloysite Nanotubes: Controlled Access and Release by Smart Gates. *Nanomaterials* **2017**, *7*, 199. [CrossRef] [PubMed]

46. Peña-Parás, L.; Sánchez-Fernández, J.; Martínez, C.; Ontiveros, J.A.; Saldívar, K.I.; Urbina, L.M.; Arias, M.J.; García-Pineda, P.; Castaños, B. Evaluation of Anti-Wear Properties of Metalworking Fluids Enhanced with Halloysite Nanotubes. *Appl. Sci.* **2017**, *7*, 1019. [CrossRef]

47. Qu, S.; Yu, Y.; Lin, K.; Liu, P.; Zheng, C.; Wang, L.; Xu, T.; Wang, Z.; Wu, H. Easy hydrothermal synthesis of multi-shelled La_2O_3 hollow spheres for lithium-ion batteries. *J. Mater. Sci. Mater. Electron.* **2018**, *29*, 1232–1237. [CrossRef]

48. Wu, G.; Wu, H.; Wang, K.; Zheng, C.; Wang, Y.; Feng, A. Facile synthesis and application of multi-shelled SnO_2 hollow spheres in lithium ion battery. *RSC Adv.* **2016**, *6*, 58069–58076. [CrossRef]

49. Shi, C.; Zhu, J.; Shen, X.; Chen, F.; Ning, F.; Zhang, H.; Long, Y.; Ning, X.; Zhao, J. Flexible inorganic membranes used as a high thermal safety separator for the lithium-ion battery. *RSC Adv.* **2018**, *8*, 4072–4077. [CrossRef]

50. Wu, H.; Wu, G.; Ren, Y.; Yang, L.; Wang, L.; Li, X. Co^{2+}/Co^{3+} ratio dependence of electromagnetic wave absorption in hierarchical $NiCo_2O_4$-$CoNiO_2$ hybrids. *J. Mater. Chem. C* **2015**, *3*, 7677–7690. [CrossRef]

51. Feng, A.; Jia, Z.; Zhao, Y.; Lv, H. Development of $Fe/Fe_3O_4@C$ composite with excellent electromagnetic absorption performance. *J. Alloys Compd.* **2018**, *745*, 547–554. [CrossRef]

52. Wang, Y.; Zhang, W.; Wu, X.; Luo, C.; Wang, Q.; Li, J.; Hu, L. Conducting polymer coated metal-organic framework nanoparticles: Facile synthesis and enhanced electromagnetic absorption properties. *Synth. Met.* **2017**, *228*, 18–24. [CrossRef]

53. Ren, X.; Liu, L.; Li, Y.; Dai, Q.; Zhang, M.; Jing, X. Facile Preparation of Gadolinium(III) Chelates Functionalized Carbon Quantum Dots-based Contrast Agent for Magnetic Resonance/Fluorescence Multimodal Imaging. *J. Mater. Chem. B* **2014**, *2*, 5541–5549. [CrossRef]
54. Wu, H.; Wu, G.; Wang, L. Peculiar porous α-Fe_2O_3, γ-Fe_2O_3 and Fe_3O_4 nanospheres: Facile synthesis and electromagnetic properties. *Powder Technol.* **2015**, *269*, 443–451. [CrossRef]
55. Ren, X.Y.; Yuan, X.X.; Wang, Y.P.; Liu, C.C.; Qin, L.; Guo, L.P.; Liu, L.H. Facile preparation of Gd^{3+} doped carbon quantum dots: Photoluminescence materials with magnetic resonance response as magnetic resonance/fluorescence bimodal probes. *Opt. Mater.* **2016**, *57*, 56–62. [CrossRef]
56. Wu, H.; Wu, G.; Ren, Y.; Li, X.; Wang, L. Multishelled metal oxide hollow spheres: Easy synthesis and formation mechanism. *Chem. A Eur. J.* **2016**, *22*, 8864–8871. [CrossRef] [PubMed]
57. Nielsen, L.E. *Mechanical Properties of Polymers and Composites*; Marcel Dekker: New York, NY, USA, 1974; Volume 2.
58. Terao, T.; Zhi, C.; Bando, Y.; Mitome, M.; Tang, C.; Golberg, D. Alignment of Boron Nitride Nanotubes in Polymeric Composite Films for Thermal Conductivity Improvement. *J. Phys. Chem. C* **2010**, *114*, 4340–4344. [CrossRef]
59. Song, W.L.; Wang, W.; Veca, L.M.; Kong, C.Y.; Cao, M.S.; Wang, P.; Meziani, M.J.; Qian, H.; LeCroy, G.E.; Cao, L.; et al. Polymer/carbon nanocomposites for enhanced thermal transport properties-carbon nanotubes versus graphene sheets as nanoscale fillers. *J. Mater. Chem.* **2012**, *22*, 17133–17139. [CrossRef]

nanomaterials

MDPI

Article

The Effect of Boron Nitride on the Thermal and Mechanical Properties of Poly(3-hydroxybutyrate-co-3-hydroxyvalerate)

Mualla Öner [1,*], Gülnur Kızıl [1], Gülşah Keskin [1], Celine Pochat-Bohatier [2] and Mikhael Bechelany [2,*]

[1] Chemical Engineering Department, Chemical-Metallurgical Faculty, Yildiz Technical University, Istanbul 34210, Turkey; kizilgulnur@gmail.com (G.K.); gulsahkeskin9@gmail.com (G.K.)
[2] Institut Européen des Membranes, IEM UMR-5635, ENCSM, CNRS, Université de Montpellier, ENSCM, CNRS, Place Eugéne Bataillon, 34000 Montpellier, France; celine.pochat@umontpellier.fr
[*] Correspondence: muallaoner@gmail.com or oner@yildiz.edu.tr (M.Ö.); Mikhael.bechelany@umontpellier.fr (M.B.); Tel.: +90-212-383-2740 (M.Ö.); +33-4-6714-9167 (M.B.)

Received: 17 October 2018; Accepted: 13 November 2018; Published: 15 November 2018

Abstract: The thermal and mechanical properties of poly(3-hydroxybutyrate-co-3-hydroxyvalerate, PHBV) composites filled with boron nitride (BN) particles with two different sizes and shapes were studied by scanning electron microscopy (SEM), differential scanning calorimetry (DSC), X-ray diffraction (XRD), Fourier Transform Infrared Spectroscopy (FTIR), thermal gravimetric analysis (TGA) and mechanical testing. The biocomposites were produced by melt extrusion of PHBV with untreated BN and surface-treated BN particles. Thermogravimetric analysis (TGA) showed that the thermal stability of the composites was higher than that of neat PHBV while the effect of the different shapes and sizes of the particles on the thermal stability was insignificant. DSC analysis showed that the crystallinity of the PHBV was not affected significantly by the change in filler concentration and the type of the BN nanoparticle but decreasing of the crystallinity of PHBV/BN composites was observed at higher loadings. BN particles treated with silane coupling agent yielded nanocomposites characterized by good mechanical performance. The results demonstrate that mechanical properties of the composites were found to increase more for the silanized flake type BN (OSFBN) compared to silanized hexagonal disk type BN (OSBN). The highest Young's modulus was obtained for the nanocomposite sample containing 1 wt.% OSFBN, for which increase of Young's modulus up to 19% was observed in comparison to the neat PHBV. The Halpin–Tsai and Hui–Shia models were used to evaluate the effect of reinforcement by BN particles on the elastic modulus of the composites. Micromechanical models for initial composite stiffness showed good correlation with experimental values.

Keywords: biopolymer; bionanocomposite; poly(3-hydroxybutyrate-co-3-hydroxyvalerate); boron nitride; mechanical properties; thermal properties

1. Introduction

Plastics are the preferred materials in many areas of daily life since they are easily processed, low-cost, light, and durable. Approximately 7% of the world's oil and natural gas is used for the production of plastics. Beside limited fossil resources, widespread usage of these nondegradable materials leads to very serious environmental problems [1,2]. The development of commercially viable bioplastics is an attractive alternative to nondegradable polymers with their renewable resources and biodegradability [3]. Polyhydroxyalkanoates (PHAs) came into prominence among biodegradable polymers synthesized by many bacteria as intracellular carbon and energy storage granules. They have very good properties such as versatility, high biodegradability in proper environmental conditions, and similar mechanical performance with petroleum-based polymers such as polypropylene (PP) [4].

Poly(3-hydroxybutyrate) (PHB) is a linear bacterial polyester. It is among the most known and the best characterized member of PHA family [5]. It is a highly crystalline polyester, is very brittle and has a very low biodegradation rate because of its high crystallinity (>90%). One solution to this problem is to copolymerize PHB with 3-hydroxyvalerate (HV) monomers in the bacterial fermentation process to form poly(3-hydroxybutyrate-co-3-hydroxyvalerate) (PHBV). The addition of HV units improves the mechanical properties, increases the thermal stability and prevents the degradation during processing. PHBVs can be obtained with different properties depending on the percentage of HB and HV units. The excess of HV units leads to the polymer being in a softer and ductile form and having a lower crystallinity value [6,7]. PHBV has a high potential application in many areas such as medical and agriculture fields and as packaging material. However, widespread applications of PHBV are still hindered by several material drawbacks such as high material costs, slow crystallization rate, poor thermal stability, brittleness and relative difficulty in processing. Recently, the addition of nanofillers and fibers as reinforcing agents into polymer matrixes to form nanocomposites has provided a promising method because they can act as nucleating agents, not only improving the polymer crystallization rates but also increasing the mechanical, thermal and/or barrier performances of the composites. Nanofillers are found to be preferable in many applications due to their high surface area-to-volume ratios and low concentrations needed to achieve reinforcing effects [8–10].

PHBV is a highly crystalline polyester with low degree of heterogeneous nucleation density. It is not thermally stable around the melting point. Molecular weight reduction can be observed during the melting process. In order to enhance the performance of PHBV the effect of various nanofillers on crystallization and mechanical behaviors has been investigated [11]. Chen et al. worked on structural and mechanical properties of PHBV/OMMT (organo-montmorillonite) nanocomposites [12]. The tensile stress of nanocomposites improved up to 32% for 3% (wt) OMMT content compared with the neat polymer. Optimum mechanical properties were obtained for 3% (wt) OMMT content. Further increasing of filler addition leads to agglomeration of OMMT causing the decrease of mechanical properties of composites [12]. Ten et al. studied the thermal and mechanical properties of PHBV/cellulose nanowhiskers biocomposites [13]. The tensile strengths of the composites have increased due to the strong interface linkages between the nanoparticles and the polymer. The tensile strength of 5 wt.% cellulose nanocomposite improved by 35.5% compared to pure PHBV, while the toughness value increased by 41% [13]. Choi et al. produced cloisite 30B organic clay-filled PHBV nanocomposites by the solution intercalation method [14]. Compared to neat PHBV, improvements in mechanical properties were observed in nanocomposites with 1, 2% and 3 wt.% cloisite 30B content [14]. Öner et al. worked on PHBV/hydroxyapatite (HAP) composites [15]. The mechanical properties of PHBV were improved using HAP particles. Xiang et al. studied the mechanical properties of PHBV/green tea polyphenol (TP) composites [16]. The results showed that the elongation at break, toughness, strain and tensile stress of composite increased with TP addition when compared to pure PHBV [16]. Luo and Netravali obtained green composites by using PHBV and pineapple fibers [17]. The flexural strength and modulus of the obtained composites, in the longitudinal direction, increased with fiber loading [17]. Nanocellulose-reinforced PHBV was prepared by Jun et al. [18]. They investigated the effect of nanocellulose types such as cellulose nanocrystals (CNC) and cellulose nanofibrils (CNFs) on mechanical properties of PHBV. The maximum tensile modulus values were obtained for 7 wt.% CNC and CNF composites but the tensile stresses of composites are lower than the tensile stress of neat PHBV [18]. The ternary cellulose/PHBV/polylactic acid (PLA) composite was developed to compromise the 100% degradability of materials [19]. Filling PHBV/PLA blends with the ball-milled celluloses increased the stiffness when using different particle sizes and filling contents [19]. The tensile strength, flexural strength and compressive strength of the composite were improved by mixing PLA fiber with PHBV [19]. Ternary nanocomposites including cellulose nanocrystals/silver nanohybrids (CNC-Ag) and biodegradable poly(3-hydroxybutyrate-co-3-hydroxyvalerate) (PHBV) were prepared by using solution casting [20]. Compared to binary PHBV/CNC nanocomposite, the ternary nanocomposites with the highest AgNPs content, showed the largest improvement in the thermal stability, mechanical, barrier, overall migration and antibacterial properties [20].

2D nanomaterials have recently been a very active research area due to their small thicknesses, wide lateral surfaces, and weak Van der Waals interactions between layers. Wang et al. produced (PHBV)/graphene nanosheet (GNS) composites via a solution-casting method and investigated their mechanical properties. The results showed that the storage modulus of PHBV/GNS composites highly improved with GNS addition [21]. Recently, research activity has increased in the area of boron nitride (h-BN) nanomaterials. Boron nitride has a layered structure, where Van der Waal's forces hold sheets of covalently bonded boron and nitrogen atoms together. The hexagonal form of BN (h-BN) is similar to graphene which accounts for their high thermal conductivity. The hexagonal boron nitride nanosheets are of great interest due to their potential use in various real-life applications. They are popular fillers for polymers due to unique material properties [22–24]. Zhi et al. studied the fabrication of boron nitride nanosheets and production of polymeric composites with improved thermal and mechanical properties [25]. BN was exfoliated via ultrasonication and used as filler for PMMA/BN nanocomposites. The thermal expansion coefficient and glass transition temperature were reduced in composites compared with the neat polymer. These results indicated that polymer chain mobility reduced due to BN nanosheet-matrix interactions. Young's modulus of PMMA was improved by 22% and strength was increased by 11% with the addition of 0.3% (wt) BN [25]. Wattanakul et al. investigated the effect of sonication and dissipation of BN on the mechanical properties of epoxy/BN nanocomposites. Impact strength of composites improved up to 33% (v) BN content and started to decrease at 37% (v) BN content. Filler particles tend to stay as agglomerates in high filler contents. Thus, the decrease of mechanical properties could be observed in high filler contents [26]. Pradhan et al. studied the effect of BN particles on mechanical properties of starch. The tensile stress of nanocomposites improved up to 3 times for 10% BN content compared to neat starch [27]. Cheewawuttipong et al. studied on the polypropylene (PP)/BN composites. The mechanical analysis results showed that the storage modulus and loss modulus increased with BN content [28]. Sun et al. combined fused silica (FS) with BN to improve mechanical properties. The flexural strength and toughness values increased significantly with 0.5 wt.% BN addition [29]. Chitosan/Boron nitride (BN) composites were prepared by solution method with variable percentage of boron nitride loading. It was found that, the thermal stability of the chitosan/BN composites was increased in comparison to virgin chitosan [30]. Zhou et al. investigated the effect of BN nanoparticles on mechanical properties of epoxy matrices. Tensile modulus increased from 2.68 ± 0.21 GPa to 3.14 ± 0.31 GPa for 50 wt.% BN loading. However, the mechanical strength, toughness, and elongation at break (%) values decreased with increasing BN content [31]. Polyamide 6 (PA6) and BN and exfoliated BN(BNNS) composites were produced by Li et al. [32]. The tensile stress of PA6/BN composites was higher than neat PA6 but for BNNS composites, the greater enhancement was obtained due to the higher aspect ratio BNNS and interaction between the polymer and BNNS [32]. The reinforcement effect of graphene-like BN on the gelatin was investigated by Biscarat et al. [33]. The barrier properties of gelatin/BN nanocomposites have been enhanced by a factor of 500 compared to a neat gelatin.

In this study, we investigated the effect of BN on mechanical and thermal properties of PHBV. This research is in the continuity of our study on the improvement of the properties of PHBV by incorporating boron nitride particles with a twin-screw extruder so that it could be transferred to industry for large-scale production. In our previous work, h-BN were studied as the potentially interesting material for the enhancement of barrier properties of PHBV [34,35]. Based on our preliminary results, the aim of this work was to investigate the mechanical and thermal properties of PHBV composites by taking into consideration the effect of different BN nanoparticles. In order to develop this understanding, polymer nanocomposites containing boron nitride nanoparticles of two different shapes, hexagonal disk (OSBN) and nanoflakes (OSFBN) were prepared through melt processing route with different concentrations. Various techniques analyzing mechanical and thermal properties were employed to characterize the PHBV/BN nanocomposites. The best mechanical properties are obtained for the nanocomposite sample containing 1 wt.% of the silanized flake type

BN (OSFBN) for which an increase in Young's modulus up to 19% was observed in comparison to the neat PHBV. The resulting biobased and biodegradable PHBV/BN nanocomposites may find potential applications in the fields of packaging and biomedical devices.

2. Materials and Methods

2.1. Materials

PHBV, the biopolymer with 8 mol% hydroxyvalerate (HV) content was purchased from ADmajoris Company, Cublize, France under the trade name MAJ'ECO FN000HA. Two different types of hexagonal nano-sized BN were used. One of the BN's was purchased from Bortek, Eskisehir, Turkey (2.27 g/cm^3, surface area 20 m^2/g); the other BN (FBN), commercial grade (PHPP325B) (2.2 g/cm^3, surface area 60 m^2/g), from Saint-Gobain Ceramics, France. Octyltriethoxysilane (OTES) was purchased from Sigma-Aldrich, Steinheim, Germany.

2.2. Surface Modification of Boron Nitride

The surface modification of boron nitride was performed by using the silanizing agent to produce an appropriate interface between matrix and filler. BN (1.5% w/v) particles were added to the 90:10 (v/v) ethanol-water mixture and treated with an ultrasonic probe system (Sonic vibra cell VCX 750, Newtown, CT, USA) for 30 min with an amplitude of 40%. The mixture was then centrifuged (Sigma 3-16P) at 4000 rpm for 35 min. Octyltriethoxysilane (OTES) with a concentration of 2.5% (w/v) was dissolved in a 90:10 (v/v) ethanol-water mixture and the pH of the solution was adjusted to 5.0 using dilute HCl solution. The solution was stirred by an ultrasonic mixer for 2 h at room temperature for silane hydrolysis. BN particles were added to the solution and sonicated with an ultrasonic probe for 30 min with an amplitude of 40%. The solution was centrifuged at 4000 rpm for 10 min. Finally, the obtained silanized BN was dried at 110 °C in an oven for 2 h and then at 65 °C in a vacuum drier.

2.3. Preparation of Nanobiocomposites

Nanobiocomposites were prepared by melt-mixing method. Double screw extruder (D: 10 mm, L/D: 20, Rondol, UK) was used to obtain PHBV/BN nanocomposites. Before extrusion, both of the polymer and boron nitride were dried at 50 °C for an hour in a vacuum dryer to remove the moisture. Extruder temperatures from the feed zone to the endpoint have been applied 90-135-160-160-150 °C, respectively. The rotation speed of the screws is 80 rpm. Polymer nanocomposite films prepared using a hot-cold press machine (Gülnar Makine, İstanbul, Turkey). The mechanical analysis specimens were cut to 2 mm thickness and dumbbell shape in accordance with ISO 527-1BA standard. The prepared nanocomposites were given in Table 1. OS code shows the silanized samples.

Table 1. The prepared nanobiocomposites for mechanical analysis.

Sample	Boron Nitride Content (wt.%)
PHBV	-
BORTEK	
PHBV/0.5OSBN	0.5
PHBV/1BN	1
PHBV/1OSBN	1
PHBV/2OSBN	2
PHBV/3OSBN	3
SAINT GOBAIN (PHPP325B)	
PHBV/0.5OSFBN	0.5
PHBV/1FBN	1
PHBV/1OSFBN	1
PHBV/2OSFBN	2
PHBV/3OSFBN	3

2.4. Characterization of BN Nanoparticles and Composites

X-Ray diffraction analyses were collected on a PHILIPS X'pert Pro Panalytical diffractometer, Egham, Surrey, UK (2θ = 2–80°, 40 kV, 20 mA, λ = 1.54 Å) in order to investigate crystalline structure of nanocomposites. The analysis was performed at room temperature. FTIR analysis was performed by using BRUKER Alpha-P, (Coventry, UK) in the 400–4000 cm^{-1} region. Scanning Electron Microscopy (SEM) was carried out by using the instrument FEI-Philips XL 30 ESEM-FEG (Amsterdam, The Netherlands) in order to investigate the morphologies of samples and the dispersion of BN particles in composites. Particle sizes were found using ImageJ software.

2.5. Thermal Properties of PHBV/BN Nanobiocomposites

In order to investigate the thermal properties of PHBV and PHBV/BN composites, DSC measurements were performed on TA Instruments (DSC Q20 V24.11 Build 124, New Castle, DE, USA). The analysis was carried out in three steps at a heating and cooling rate of 10 °C/min in an aluminum crucible under 50 mL/min nitrogen atmosphere. In the first heating step, samples of 10 mg mass were heated from 0 °C to 200 °C at a rate of 10 °C/min and kept at this temperature for 2 min to erase thermal history of the material. Then the samples were cooled from 200 °C to 0 °C at a cooling rate of 10 °C/min (cooling run) and kept at this temperature for 2 min. Then, the samples were re-heated to 200 °C at a rate of 10 °C/min. The melting and the crystallization temperatures (T_m and T_c) as well as the melting and the crystallization enthalpies (ΔH_m and ΔH_c) were determined. The crystallinity was calculated from the formula below [34]:

$$\chi_C(\%) = \left[\frac{\Delta H_m}{\left(W_{PHBV} \times \Delta H_m^{ref} \right)} \right] \times 100 \tag{1}$$

where ΔH_m is the melting of sample, W_{PHBV} is the weight fraction of PHBV in the composite and ΔH_m^{ref} is the theoretical melting enthalpy for 100% crystallized PHBV, 146 J/g [34].

TGA analysis was performed using a thermogravimetric analyzer (TA Instruments, Q500 V 20.13 Build 39, New Castle, DE, USA). About 10 mg sample was weighed and analyzed in a platinum crucible by heating at a heating rate of 10 °C/min up to 800 °C under a 40 mL/min nitrogen-60 mL/min air environment.

2.6. Mechanical Properties of PHBV Nanobiocomposites

Uniaxial tensile testing was performed according to ASTM D882-12 standard, by using 2 kN capacity Devotrans (161070 CKS GP, Istanbul, Turkey) mechanical testing machine. Specimens were kept at 50 °C in a ventilated oven for 48 h for conditioning before the test. Mechanical analyses of nanocomposites were performed using 5 mm/min tensile rate and 1 N preload at room temperature. Tensile strength at break, Young's modulus, and elongation at break values were determined from the stress-strain curves. Five specimens of each sample group were tested, and the average results were reported.

3. Results and Discussion

3.1. Morphological Characterization of BN Particles by SEM

The morphology and the size of the BN particles were examined by scanning electron microscopy (SEM). Figure 1 compares the SEM images of silanized OSFBN and OSBN particles after 60 min ultrasonic treatment. The OSFBN particles showed predominantly irregular flake-type shaped morphology whereas, in case of the OSBN particles, the predominant shape was nearly hexagonal disk particles as shown in Figure 1a,b. We examined SEM micrographs of several samples by using Image J software and performed statistical analysis on the particle sizes. The results show that the size of the particles was reduced after ultrasonic treatment. The mean length (L), the mean width (w) and thickness (t) of the OSFBN particles were reduced from 2445.9 ± 1507.6 nm to 765.4 ± 376.8 nm; from 1483.7 ± 853.8 nm to 360.49 ± 177.8 nm and from 249.7 ± 137.4 nm to 19.0 ± 5.1 nm, respectively. The mean diameter (d) and thickness (t) of OSBN particles, reduced from 225.0 ± 108.0 nm to 163.3 ± 72.9 nm and from 61.6 ± 25.8 nm to 39.7 ± 10.7 nm, respectively after ultrasonication. The standard deviations given here describe the variation in the mean value calculated from separate SEM images. The range of values was on the order of hundreds of nanometers, indicating the polydispersity of each particle system.

Figure 1. Scanning Electron Microscopy (SEM) micrograph of particles after 60 min ultrasonication (**a**) silanized BN (**b**) silanized FBN.

3.2. SEM of Nanocomposites

The structure of the PHBV/BN nanocomposites was investigated using SEM to get a qualitative understanding of the dispersion of BNs through direct visualization. Figure 2 shows the cross-section SEM images of the cryo-fractured surfaces of 1 wt.% PHBV/OSBN and PHBV/OSFBN composite samples. The dispersion of boron nitride in composites can be seen in the SEM figures. A good dispersion of the BN particles in the matrix was observed at 1 wt% loading.

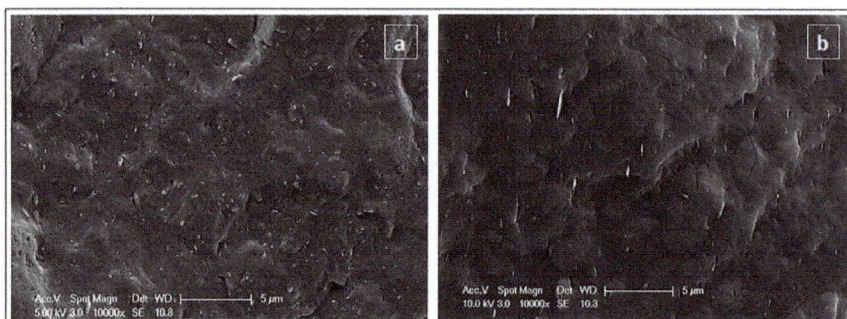

Figure 2. SEM images of composites (**a**) PHBV/1OSBN (**b**) PHBV/1OSFBN.

3.3. Characterization by XRD and FTIR

XRD analyses were carried out to investigate the crystalline structure of prepared nanocomposites (Figure 3). The XRD patterns of PHBV exhibited characteristic 2θ peaks at 13.6° (020), 17.1° (110), 19.9° (021), 21.7° (101), 22.3° (111), 25.5° (121), 27.1° (040) and 30.3° (002) [36,37]. Boron nitride exhibited characteristic 2θ peaks at 26.80° (002). Nanocomposites show same reflections as neat PHBV indicating that boron nitride incorporation did not change the unit cell and crystalline structure of PHBV. However, the intensity of (020) and (110) peaks of PHBV changes with BN content as shown in Figure 4. While the intensity of (020) peak of PHBV increases, the intensity of (110) peak of PHBV becomes lower with boron nitride addition. The increase of (020) peak of PHBV could be related to the crystallite lamella size of polymer and indicated that crystallization is promoted by boron nitride addition [38]. The decrease of (110) peak of PHBV indicates the restricted crystal growth in (110) plane. (020)/(110) relative intensity ratios for PHBV and nanocomposites were given in Table 2. While the relative intensity ratio of the neat PHBV matrix was 1.42, this value increased up to 4.58 for the OSBN nanocomposites and 6.72 for OSFBN nanocomposites. The increase in the relative ratio implies that crystal growing was promoted in (020) crystal plane as reported [38,39].

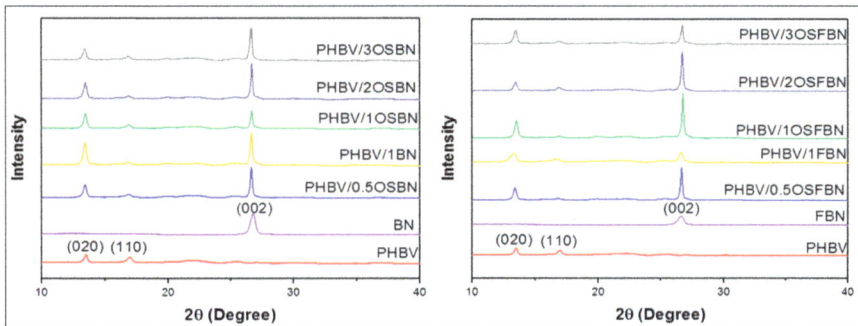

Figure 3. XRD patterns of neat PHBV, BN, FBN and nanocomposites.

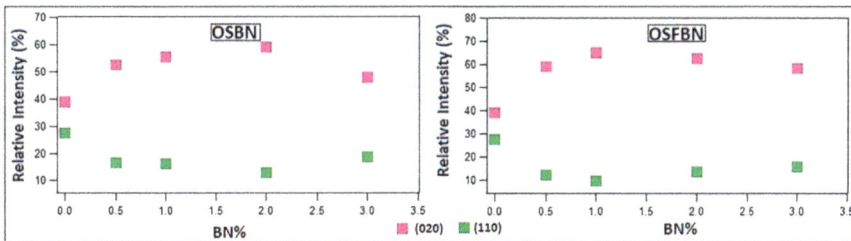

Figure 4. (020) and (110) relative intensity variation with BN content.

Table 2. (020)/(110) relative intensity ratios of nanocomposites.

(020)/(110)	
PHBV	1.42
PHBV/0.5OSBN	3.17
PHBV/1OSBN	3.40
PHBV/2OSBN	4.58
PHBV/3OSBN	2.56
PHBV/0.5OSFBN	4.98
PHBV/1OSFBN	6.72
PHBV/2OSFBN	4.70
PHBV/3OSFBN	3.76

Figure 5 shows the FTIR spectra of neat PHBV and the prepared nanocomposites. PHBV exhibited some characteristic peaks as CH_3 asymmetrical stretching at 3015–2960 cm^{-1}, CH_2 asymmetrical stretching at 2945–2925 cm^{-1}, CH_3 symmetrical stretching at 2885–2865 cm^{-1}, C=O stretching at

1723–1740 cm^{-1}, CH$_2$ wagging at 1320–1159 cm^{-1} [40], asymmetrical –C–O–C– stretching, symmetrical –C–O–C– stretching at 800–975 cm^{-1} [41], CH$_2$ scissoring at 1453–1459 cm^{-1}, C–O stretching at 1065–1030 cm^{-1} and C–C stretching at 979–980 cm^{-1} [42]. BN exhibited characteristic peaks as B–N at 1300–1400 cm^{-1} and B–N–B at 775–820 cm^{-1} [43,44]. OTES exhibited Si–O stretching at 1053–1114 cm^{-1} and CH$_n$ (C–H) stretching at 2850–3000 cm^{-1} [45,46]. The silane peaks were not observed in FTIR spectrum of composites because of overlapping of silane peaks with PHBV peaks in the same region. Figure 6 shows a comparison between the infrared spectrum of silanized and nonsilanized BN particles. After both BN particles were treated with silane, the spectrum showed new bands in addition to the characteristic peaks of BN. In the spectrum of OSBN and OSFBN, the bands at 2850–3000 cm^{-1} regions were attributed to the CH$_2$ asymmetric and symmetric stretching vibration, respectively, which originated from the silane-containing molecule. The band at 1053 cm^{-1} is assigned to the in-plane Si–O stretching and that at 1114 cm^{-1} is assigned to the perpendicular Si–O stretching.

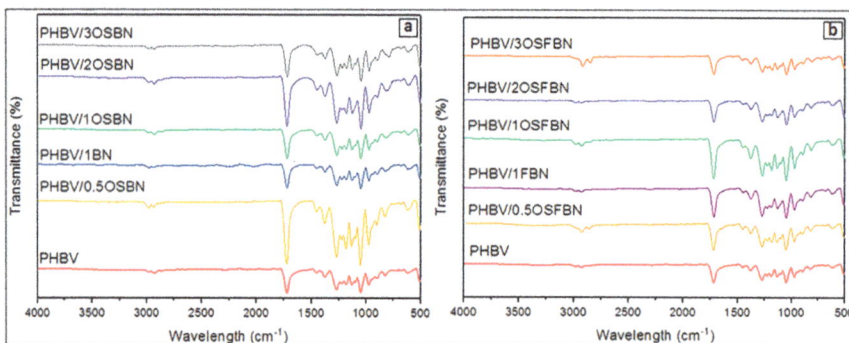

Figure 5. FTIR spectrum of nanocomposites (**a**) PHBV/BN (**b**) PHBV/FBN.

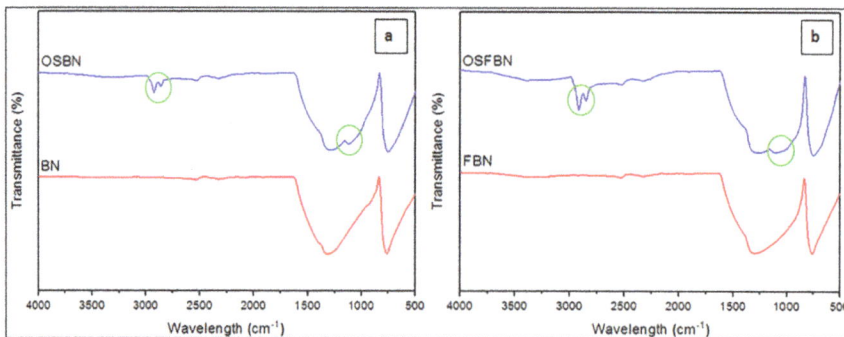

Figure 6. FTIR peaks of silanized (**a**) BN and (**b**) FBN.

3.4. Thermal Stability of Nanocomposites

The thermal stability of the PHBV and PHBV/BN composites were studied using Thermo Gravimetric Analysis (TGA) to measure the degradation temperature of different samples. The TGA thermograms for neat PHBV was compared with that of the PHBV/BN nanocomposites. TGA curves of PHBV and composite samples are represented in Figure 7. It is noted that weight loss of PHBV and its composites occurs in a one-step process between 230 °C and 300 °C. No significant weight loss was recorded before 200 °C for all samples. After 200 °C, weight loss proceeds very rapidly and the polymer completely degrades by 300 °C. It has been established that the thermal degradation of PHBV was due to the rupture of ester bonds during chain scission [47]. This could be attributed to PHBV copolymer separating out into individual PHV and PHB units. The temperatures of at 10% weight loss (T_{10}), at 50% weight loss (T_{50}), the initial decomposition temperature (T_i), and the maximum rate of degradation temperature (T_{max}) are presented in Table 3. As shown in Table 3, T_i, T_{10}, T_{50} and T_{max} values increased in the composites. When the initial weight loss is taken as a point of comparison, the onset degradation temperature (T_i) for neat PHBV is 234.45 °C and increases to 252.70 °C and 251.10 °C for PHBV/1OSBN and PHBV/1OSFBN composites. The surface treatment by silane results in improved initial thermal stability in comparison to untreated BN. Another important thermal property is the temperature corresponding to the maximum rate of weight loss (T_{max}). T_{max} shifted to higher temperatures as the BN content increased, from 275.08 °C to 295.50 °C. This result showed that the thermal stability of the composites improved with the addition of the BN to the polymer matrix. One of the most important property of boron nitride is its high-temperature resistance. As a result of this property, inclusion of the BN within a polymeric matrix results in increasing of the thermal stability of the composite. The fabrication of gelatin-BN nanocomposites was investigated by Biscarat et al. [33]. An increase of gelatin degradation temperature was observed by using DSC. It was concluded that gelatin chains that intercalate into BN are restricted by the nanosheets, and the movement of segments is restrained. An electrostatic interaction or a hydrogen bond between the charged groups of gelatin chains and BN acts as physical crosslinking and reduces the activity of the gelatin [33].

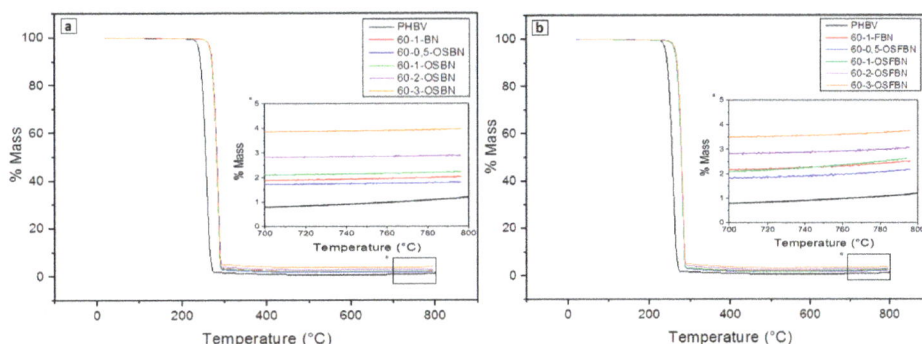

Figure 7. TG thermograms of (**a**) PHBV/OSBN and (**b**) PHBV/OSFBN nanocomposites.

Table 3. TG results of nanocomposites.

Sample	T_i (°C)	T_{10} (°C)	T_{50} (°C)	T_{max} (°C)	Char (%)
PHBV	234.45	243.50	256.04	275.08	1.81
PHBV/0.5OSBN	251.90	271.35	283.39	293.90	1.82
PHBV/1BN	250.30	271.97	282.97	292.30	2.03
PHBV/1OSBN	252.70	270.67	282.10	294.70	2.23
PHBV/2OSBN	253.50	271.53	282.73	295.50	2.88
PHBV/3OSBN	254.30	271.06	282.30	295.50	3.97
PHBV/0.5OSFBN	248.92	269.00	279.45	289.44	2.17
PHBV/1FBN	247.09	268.18	278.92	289.54	2.53
PHBV/1OSFBN	251.10	267.82	278.37	289.90	2.62
PHBV/2OSFBN	251.90	269.40	279.50	291.06	3.06
PHBV/3OSFBN	253.45	270.17	280.11	294.17	3.75

Camargo et al. investigated thermal behavior of PHBV/Lignin composites by thermogravimetric analysis and found that the thermal decomposition of pure PHBV and composites took place in a single degradation step [48]. In another study, Bhardwaj et al. worked on the thermogravimetric analysis of PHBV/cellulose fibers [49]. Cellulose fibers did not affect the existing degradation step of PHBV. Lai et al. studied the thermal properties of multilayered carbon nanotube/PHBV composites [50]. It was observed that the degradation temperature rise was up to 16 °C for the PHBV nanocomposite. It was concluded that nanodispersion of carbon nanotubes increased thermal stability of composites [50].

3.5. Thermal Properties of Composites

Differential scanning calorimetry (DSC) was used to study the change in enthalpy values associated with chemical phase transitions in composite samples, as a function of temperature. Two heating and one cooling cycles were performed in order to make useful comparisons for PHBV/OSFBN and PHBV/OSBN composites. Figure 8 and Table 4 present the thermograms obtained from the cooling, first and second heating cycle at 10 °C/min for PHBV and PHBV/BN composites. The various thermal property results calculated from the DSC heating and cooling curves are summarized in Table 4. As can be observed from Figure 8, neat PHBV and PHBV/BN composites showed two melting peaks during the first heating. The peak maximum temperatures of first and the second melting peaks are given in Table 4 as T_{m1} and T_{m2}. The double melting endotherms have been reported by several groups [51–53]. The double melting peak in polymers may be due to several reasons. The origin for the double melting behavior of PHBV is still being researched. It was generally accepted that the double melting peaks were caused by melting–recrystallization–melting behavior during heating scans [52]. The first melting peak values are in the range of 166–171 °C for composites as opposite to 170 °C for neat PHBV. The degree of crystallinity (Xc) from the first heating scan was computed and presented in Table 4. The BN addition does not influence crystallinity of the matrix for silanized samples up to 3 wt% loadings. The addition of the surface treated BN to the polymer matrix (PHBV/1OSBN) slightly increased the crystallinity of PHBV from 57% to 60%.

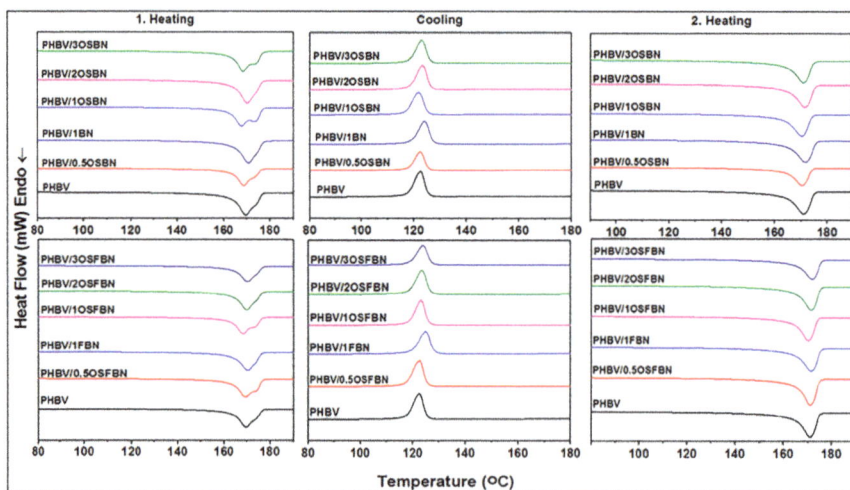

Figure 8. DSC thermograms of PHBV/BN and PHBV/FBN nanocomposites.

Table 4. DSC results of nanocomposites.

Sample	First Heating				Cooling		Second Heating		
	T_{m1} (°C)	T_{m2} (°C)	ΔH_{m1} (j/g)	X_c (%)	T_{c1} (°C)	ΔH_c (j/g)	T_{m1} (°C)	ΔH_{m2} (j/g)	X_c (%)
PHBV	170	173	88	60	122	86	171	100	68
PHBV/0.5OSBN	169	174	87	60	122	88	170	98	67
PHBV/1BN	171	-	81	57	124	84	172	95	65
PHBV/1OSBN	166	173	86	60	121	91	168	100	69
PHBV/2OSBN	171	-	86	60	122	87	170	100	70
PHBV/3OSBN	168	175	83	56	123	81	164	90	64
PHBV/0.5OSFBN	170	175	87	60	123	89	171	99	68
PHBV/1FBN	171	-	87	60	125	89	171	100	69
PHBV/1OSFBN	169	173	89	61	122	90	169	102	71
PHBV/2OSFBN	170	-	85	59	124	84	172	95	67
PHBV/3OSFBN	171	-	82	58	124	81	172	91	65

The thermograms and data obtained from the second heating cycle provide information on the crystallization and melting behavior of the samples without the influence of different thermal histories. The broad or double melting peaks observed during the first heating cycle were not present in the second heating cycle. Table 4 shows the degree of crystallinity (%) of PHBV and PHBV/BN composites measured during the second heating cycle. In general, it was observed that the degree of crystallinity was higher when measured during the second heating cycle compared with that measured during the first heating cycle. The addition of BN increased the degree of PHBV crystallinity from 68% to 70% and 71% for PHBV/2OSBN and PHBV/1OSFBN composites respectively. Further increasing of the BN concentration to 3% decreased the crystallinity values of the composites. This behavior can be explained by the particle dispersion and distribution in the PHBV matrix. When the particles are dispersed well, the higher surface area helps the particles to act as nucleating agent. The change in particle size due to aggregation and the surface characteristics of the particle leads to the decrease in crystallinity. Similar to the first heating cycle, the addition of treated BN increased the crystallinity of PHBV. The crystallinity of PHBV/BN composites treated with silane was slightly higher than that displayed by composites without any silane treatment except for sample PHBV/3OSBN.

The crystallization temperature (T_{c1}) and the heat of crystallization (ΔH_c) were determined from the DSC cooling runs of samples. These are shown in Table 4 for composites with different BN contents. From Table 4, cold crystallization temperature nanocomposites did not much change with the change in nanoparticle concentration. The heat of crystallization of 1 wt.% PHBV/1OSBN and PHBV/1OSFBN composites was higher than the neat PHBV, while it decreased with the increase in nanoparticle loading. This behavior can be explained by agglomeration of BN particles at higher loadings.

In a study by Sanchez Garcia et al., Tm values of PHBV nanocomposites did not change or reduced slightly [54]. There are many factors that affect the melting temperature, such as molecular order, crystal thickness, and crystal perfection. Polymer degradation due to melt processes and chain separation can cause this small decrease, but other factors are also influential [54]. Hassaini et al. investigated thermal properties of PHBV/olive husk flour (OHF) composites by DSC [55]. They found that the melting temperatures of PHBV did not change with OHF addition, however, the melting enthalpies increased. It shows that OHF filler contributes to the crystallization of PHBV as a nucleating agent. The effect of graphene nanoplatelets (GNPs) on the mechanical properties of high-density polyethylene (HDPE) nanocomposites was investigated [56]. The crystallinity of the HDPE composite was decreased by increasing the concentration of GNPs as a result of the formation of the smaller crystalline domains in HDPE in the presence of nanomaterials. Crystal defects in the presence of inhomogeneities were believed to decrease the matrix crystallinity [56]. Covalent bonds across the interface have been shown to increase crystallinity whereas attractive noncovalent interactions have shown decreasing or unchanging in crystallinity in comparison to the neat polymer [57,58]. Yu et al. reported that the crystallinity of the electrospun fibers of PHBV decreases when adding ZnO [59].

This result was explained by the decrease of the PHBV crystallinity decreased because of the formation of hydrogen bonds between PHBV and ZnO.

3.6. Mechanical Analysis Results

There are many factors that influence the mechanical properties of nanocomposites. The vast majority of these factors depend on the properties of the fillers such as size and aspect ratio, distribution and orientation of the particles. The main difference between traditional fillers and nanofillers is that nanosize materials have a much wider surface (interface) per unit volume [60]. Tensile tests are applied in order to investigate the effect of BN addition and silanization on mechanical properties. Properties such as tensile modulus, tensile strength, and strain-at-break were measured as shown in Figures 9 and 10. As shown in Figure 9, adding 0.5 wt% OSBN essentially had no effect on the modulus. PHBV with 2 wt% OSBN had the highest value of modulus. It increased as well the neat PHBV stiffness by 8%. The addition of 3 wt% OSBN decreased the composite Young's modulus and the tensile strength. The addition of OSBN did not have a significant impact on the tensile strength of the composites. Furthermore, the elongation at break of neat PHBV decreased with increased OSBN loading. Maximum elongation decreased, from 2.1% for neat PHBV to 1.9% with the addition of 3 wt% OSBN. The slight increase in strength and the decrease in elongation in the composites might be attributed to an embrittlement caused by BN agglomeration.

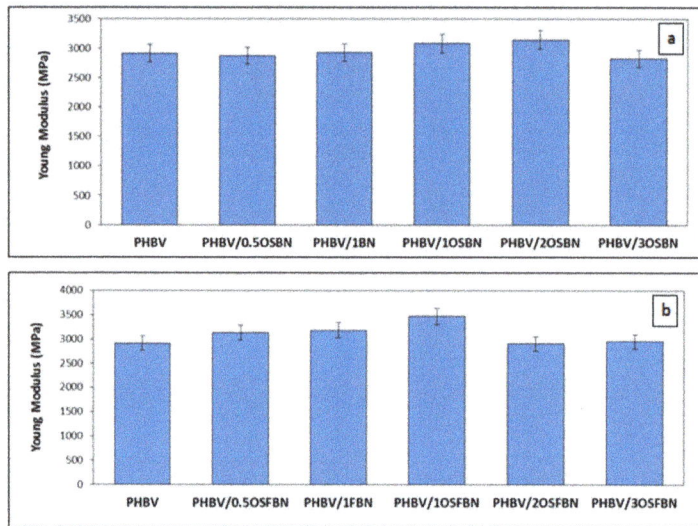

Figure 9. Young's modulus values of (**a**) PHBV/BN and (**b**) PHBV/FBN nanocomposites.

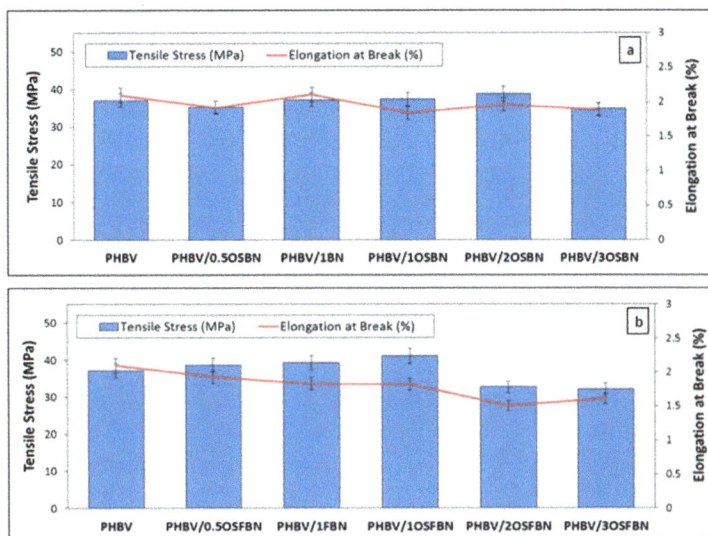

Figure 10. Tensile stress and elongation at break values of (**a**) PHBV/BN and (**b**) PHBV/FBN nanocomposites.

Both the Young's modulus and tensile strength of PHBV/OSFBN composites were increased for films containing 1 wt% OSFBN content. The addition of nanocrystals caused enhancement of the Youngs modulus and the tensile strength up to 1 wt% but led to a decrease at higher loadings. Young's modulus for the PHBV/1OSFBN composite was found to be around 3469.7 MPa which accounts for a maximum 19% increase. The maximum increase of the tensile strength of PHBV/1OSFBN composite is 10.6%. The elongation at break of PHBV decreased with the addition of OSFBN. This result shows that BN filler particles were well dispersed at low filler loadings but nonhomogenously distributed at higher concentrations.

These findings are in agreement with the XRD results. Figure 11 shows the Young's modulus of the composites as a function of (020)/(110) relative intensity ratio for neat PHBV and PHBV/OSBN and PHBV/OSFBN nanocomposites. As seen from the Figure 11, all composites had higher (020)/(110) ratio than the neat PHBV. 1 wt.% PHBV/OSFBN composite samples showed the highest (020)/(110) ratio and the highest Young's modulus compared to other composites as well as neat PHBV. This was consistent with the decrease in the (020)/(110) ratio of 3 wt.% PHBV/BN samples. Young's modulus of 3 wt.% PHBV/BN composite samples showed the lowest (020)/(110) ratio. This result associated to XRD analyses shows that BN filler particles at low filler loadings directly influence the morphological organization of PHBV polymer matrix and increase the stiffness of the polymer. However, the effect of nanoparticles addition is not linear and above a certain limit there is a reduction of stiffness enhancement. Maximizing the properties of the polymer at low nanoscale in loadings of a nanoscale filler requires that it is thoroughly distributed throughout the polymer matrix and that complete exfoliation of the filler's layers has occurred.

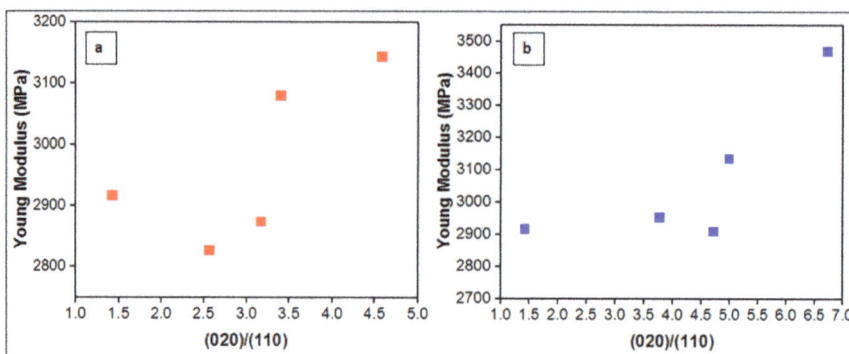

Figure 11. Young's modulus variation with (020)/(110) relative intensity ratio of (**a**) PHBV/OSBN and (**b**) PHBV/OSFBN composites

For the same BN content, the OSFBN composites had higher tensile strength and modulus than OSBN composites. These results showed that the mechanical properties of the composites were found to increase more for the PHBV/OSFBN compared with PHBV/OSBN samples. The higher values in strength and modulus observed in the PHBV/OSFBN specimen compared with the PHBV/OSBN can be attributed to different factors: (1) OSFBN was dispersed more uniformly in the composite specimens than the OSBN; (2) this may be related to the higher surface area displayed by OSFBN ($71.90 \text{ m}^2/\text{g}$), which can promote a better intercalation of BN nanosheets between PHBV chains in comparison to OSBN ($26.89 \text{ m}^2/\text{g}$) and can influence the polymer chain organization. Studies concerning surface area have shown that reinforcement is related to nanoparticle surface area [61]. The larger surface area of OSFBN leading to the stronger interactions between the BN and PHBV. In this work, the partial exfoliation of the OSFBN allows for maximum surface area exposure between the filler and PHBV; (3) the flake-like nanoparticles showed higher tensile strength than disk type particles.

It is well known that the homogeneous dispersion of nanoparticles in the polymer matrix is necessary to improve properties of composite. The good dispersion of the both BN particles in the matrix was observed in SEM figures (Figure 1a,b). Apart from the distribution of the filler, understanding the effect of the shape and aspect ratio of the filler on the composite's properties impacts its design. As we have observed in SEM images, the aspect ratio for OSFBN flakes are around 19 and OSBN disks be around 4. The property improvement of the OSFBN may be attributed the higher surface area and aspect ratio of the platelets.

To obtain a strong interface, the filler should have an attractive interaction with the matrix. The interaction may be achieved through surface chemistry in the form of functionalization with chemical moieties. In order to see surface functionalization on mechanical properties, the composite samples were prepared without using coupling agent. Young' modulus values of nonsilanized nanocomposites with 1 wt.% BN and 1 wt.% FBN content are 2929 MPa and 3187 MPa respectively. After silanization, the modulus values increased to 3080 MPa and 3469 MPa, for OSBN and OSFBN nanocomposites respectively. This shows that good interfacial adhesion was achieved between PHBV and BN, which might be due to the silane coupling agent used in this study.

Carotenuto et al. prepared LDPE/GNP (low-density polyethylene/graphite nanoplatelets) nanocomposites and tested the mechanical performance. It was found that both tensile elongation at break and compressive extension were reduced in films containing GNP [62]. It was claimed that this is due to the obstruction in polymer chain mobility, where the polymer chains are not allowed to unfold and rotate when stress is applied due to the uniform presence of GNP throughout the LDPE matrix. Yu et al. prepared PHBV/cellulose nanocrystal-silver (CNC-Ag) nanocomposites and tested the mechanical performance of nanocomposites [36]. They obtained the highest Young's modulus and the lowest elongation at break for 10% CNC-Ag content. These improvements have been shown to be a consequence of the homogeneous distribution of the CNC-Ag additive, the increased interfacial adhesion between two phases by means of interaction of hydrogen bonds, and the increased crystallization of PHBV. However, with further increase of CNC-Ag filler content to 15 wt.% or more let to the decrease in mechanical properties was observed. This result was explained by the agglomeration of the filler. Xiang et al. prepared PHBV/tannic acid nanocomposites and examined the mechanical performance. Mechanical properties such as tensile stress and fracture toughness were improved comparing to neat PHBV matrix. However, no improvement was observed above a certain level of filler content [63].

3.7. Mechanical Modelling

Multiple analytical, mechanics-based theories were developed to model particles-filled composite structures. These theories are reliant on filler volume fraction and elastic properties of each constituent. As a result, these models offer a good indication of the resultant properties but are unable to account for the effect of factors, such as particles interaction and distribution, on the properties of the composites. All models used for the calculation of relative modulus assumed perfect interfacial adhesion between particles and matrix. In this work, the experimental data obtained were compared with two of the simplest and most common theoretical models to predict Young's modulus of the composite materials; Halpin–Tsai and Hui–Shia models. The Halpin–Tsai model accounts the modulus of reinforcement and matrix materials as well as shape and volume fraction of filler. The mechanical modeling of a variety of reinforcement of fillers such as platelet-like or flake-like fillers can be done by using this model [64]. The Halpin–Tsai equation for randomly oriented discontinuous fillers is given in Table 5 [65]. The Hui–Shia model is employed to the mechanical modeling of composites with the assumption of perfect interfacial bonding between the polymer matrix and fillers. The Hui–Shia equation for platelet fillers is given in Table 5. In the Hui–Shia model, α represents the aspect ratio of the filler, which is the width of the platelet divided by the thickness of particles. In this work, SEM images were used for calculation of aspect ratio. As observed by SEM, the shape of the OSBN nanoparticles was a nearly hexagonal disk with an aspect ratio of approximately 4.1. The OSFBN nanoparticles showed irregular flake-like shapes with the mean aspect ratio of 18.9. E_c, E_m and E_f symbolize the Young's modulus of nanocomposite, matrix and filler respectively. E_m value was taken from directly Admajoris as 2.95 GPa. E_f, the Young's modulus of hexagonal boron nitride was taken from literature as 40 GPa [24]. The factors η_L and η_T are given by equations in Table 5 as a function of E_f (the modulus of the filler) and E_m (modulus of the matrix). Φ_f symbolizes the volume fraction of nanoparticle in composite, it is defined as:

$$\phi_f = \frac{V_{BN}}{V_{BN} + V_{PHBV}} \tag{2}$$

Table 5. Mechanical models [64,65].

Model	Array Type	Formula
Halpin–Tsai Model	Random array	$\frac{E_c}{E_m} = \frac{3}{8}\left(\frac{1+\zeta\eta_L\phi_f}{1-\eta_L\phi_f}\right) + \frac{5}{8}\left(\frac{1+2\eta_T\phi_f}{1-\eta_T\phi_f}\right)$ $\eta_L = \frac{\left(\frac{E_f}{E_m}\right)-1}{\left(\frac{E_f}{E_m}\right)+\zeta}$ $\eta_T = \frac{\left(\frac{E_f}{E_m}\right)-1}{\left(\frac{E_f}{E_m}\right)+2}$
Hui–Shia Model	Regular array	$\frac{E_c}{E_m} = \frac{1}{1-\frac{\phi_f}{4}\left(\frac{1}{\xi}+\frac{3}{\xi+\Lambda}\right)}$ $\xi = \phi_f + \frac{E_m}{E_f-E_m} + 3\left(-\phi_f\right)\left(\frac{(1-g)\alpha^2-\frac{g}{2}}{\alpha^2-1}\right)$ $g = \frac{\pi}{2}\alpha$ $\Lambda = \left(1-\phi_f\right)\left(\frac{3(\alpha^2+0.25)g-2\alpha^2}{\alpha^2-1}\right)$

The volume fraction for composites with 0.5 wt.%, 1 wt.%, 2 wt.% and 3 wt.% BN content was calculated using this formula. In the Halpin–Tsai model, ζ which depends on the width (w) and thickness (t) of nanoparticles and symbolizes the shape factor is defined as:

$$\zeta = 2\left(\frac{w}{t}\right) \tag{3}$$

In the Hui–Shia model, g is the geometrical parameter which depends on the aspect ratio of the filler. The parameters ζ and Λ are defined in equations given in Table 5.

Figure 12 compares the model results predicted by the Halpin–Tsai and Hui–Shia empirical relations for the modulus of the PHBV/BN composites with the experimental results obtained from the tensile testing. Generally, theoretical modeling was in good agreement with experimental data. From the figure, the measured values of PHBV/OSFBN samples at low loadings are higher than predicted value of the models, but model equations displayed higher modulus values than the obtained experimental values at higher loadings. The models overestimate reinforcement phenomena at higher loadings. This may be due to several causes; the presence of agglomerations in the matrix weakening the structure, the inhomogeneous distribution of the BN reinforcement by extrusion in higher loadings, and the weak interaction between the nonpolar filler and the nonpolar matrix. Improvements in any of these three facets should result in moduli that are more accurately predicted by the models. On the other hand, both models are heavily reliant on filler volume fraction of the reinforcing phase. It should be noted that for rigid reinforcing phase, an increase in particle volume fraction often correlates to an increase in theoretical Young's modulus. Both models have been shown to underestimate the potential of particles to reinforce polymers at lower loadings. This discrepancy may be attributed to the lack of model's ability to predict reinforcement effect of particles at low loadings. The deviation (%) in E_c/E_m of nanocomposites for different mechanical models is given in Table 6. Model predictions were in good agreement with experimental results.

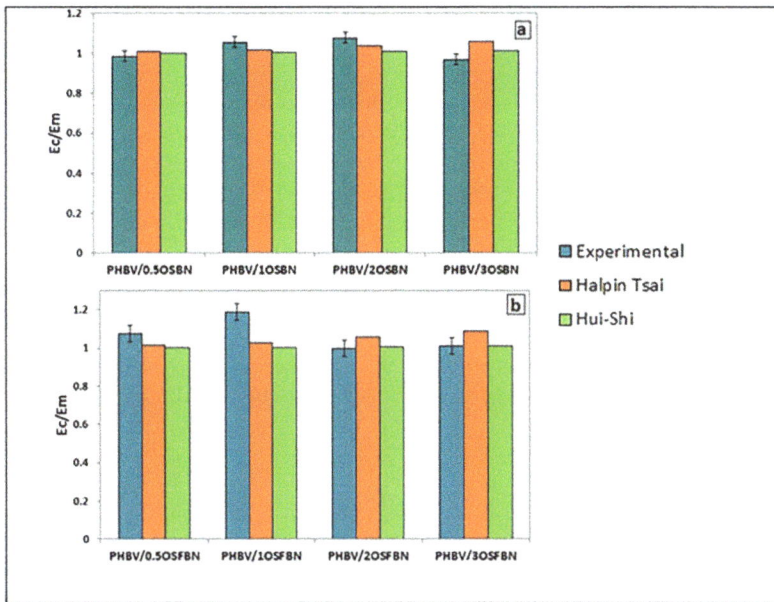

Figure 12. Comparison of mechanical models with experimental data for (**a**) PHBV/OSBN and (**b**) PHBV/OSFBN nanocomposites.

Table 6. Deviation (%) in E_c/E_m of nanocomposites for mechanical models.

Sample	Halpin–Tsai Deviation (%)	Hui–Shia Deviation (%)
PHBV/0.5OSBN	2.480	1.743
PHBV/1OSBN	3.447	4.829
PHBV/2OSBN	3.575	6.306
PHBV/3OSBN	9.353	4.757
PHBV/0.5OSFBN	5.578	6.730
PHBV/1OSFBN	13.473	15.560
PHBV/2OSFBN	6.133	1.123
PHBV/3OSFBN	7.460	0.011

4. Conclusions

PHBV nanocomposites containing BN nanoparticles with different sizes, shapes, and specific surface areas were processed and characterized to elucidate the effects of nanoparticles on nanocomposite properties. PHBV/BN nanocomposite processing which is scalable from the laboratory to an industrial setting was prepared via a masterbatch by using the twin-screw extruder. SEM morphology showed the good dispersion of BN in the PHBV matrix with using coupling agent at low loadings. The interaction between PHBV and boron nitride was evidenced by FTIR and XRD. TGA analysis showed that the thermal decomposition of PHBV was retarded by the interaction between BN and PHBV. DSC measurements did not show significant alterations to the crystalline content of PHBV. The matrix crystallinity was not significantly affected by the change in filler concentration but decreasing of the crystallinity of PHBV/BN composites was observed in higher loadings. The crystallinity of the PHBV was not affected by the differences in the nanoparticle shapes which indicated that the interfacial interactions between polymer and both particle systems were weakly attractive. The nanoparticles were likely to be located in the amorphous regions of the PHBV matrix as the crystalline content of polymer matrices did not change significantly with the increase in particle concentration.

BN increased the modulus of PHBV, as observed via tensile tests. However, elongation at break decreased as the amount of BN increased. The Young's modulus of the composite was increased by 19% in PHBV/1OSFBN composite. At higher loading, BN agglomeration takes place within the PHBV matrix and the lack of proper adhesion between the matrix and the BN results in insufficient stress transfer. Moreover, modified BN has been shown to contribute to an enhancement of Young's modulus than unmodified BN. The differences in the mechanical behavior of the two BN particles were attributed to the different surface area and aspect ratio characteristics. The higher surface area $(71.9 \text{ m}^2/\text{g})$ and aspect ratio (18.9) displayed by OSFBN can promote a better intercalation of BN nanosheets between PHBV chains and influence the mechanical properties. Theoretical modeling can be used to predict the potential modulus improvements of adding a filler to a matrix. In our research, The Halpin–Tsai and Hui–Shia models were used to evaluate the effect of reinforcement by BN particles on the elastic modulus of the resulting composites. Both model equations predict values close to the experimental results.

Author Contributions: M.Ö., M.B. and C.P.-B. conceived and designed the experiments; G.K. (Gülnur Kızıl) and G.K. (Gülşah Keskin) performed the experiments; M.Ö., M.B., C.P., G.K. (Gülnur Kızıl) and G.K. (Gülşah Keskin) analyzed the data; M.Ö. wrote the paper.

Funding: This research was funded by Scientific and Technological Research Council of Turkey (TÜBİTAK, Project No: 215M355), and the Campus France (PHC Bosphore No: 35211) under a Bilateral Cooperation Program between Turkey and France. G.K. (Gülşah Keskin) and G.K. (Gülnur Kızıl) gratefully acknowledge TÜBİTAK for the scholarship.

Conflicts of Interest: The authors declare no conflicts of interest.

References

1. Malmir, S.; Montero, B.; Rico, M.; Barral, L.; Bouza, R.; Farrag, Y. Effects of poly(3-hydroxybutyrate -co-hydroxyvalerate) microparticles on morphological, mechanical, thermal, and barrier properties in thermoplastic potato starch films. *Carbohydr. Polym.* **2018**, *194*, 357–364. [CrossRef] [PubMed]
2. Williams, C.K.; Hillmyer, M.A. Polymers from renewable resources: A perspective for a special issue of polymer reviews. *Polym. Rev.* **2008**, *48*, 1–10. [CrossRef]
3. Bordes, P.; Pollet, E.; Avérous, L. Nano-biocomposites: Biodegradable polyester/nanoclay systems. *Prog. Polym. Sci.* **2009**, *34*, 125–155. [CrossRef]
4. Kushwah, B.S.; Kushwah, A.V.S.; Singh, V. Towards understanding polyhydroxyalkanoates and their use. *J. Polym. Res.* **2016**, *23*, 1–14. [CrossRef]
5. Poirier, Y.; Nawrath, C.; Somerville, C. Production of polyhydroxyalkanoates, a family of biodegradable plastics and elastomers, in bacteria and plants. *Nat. Biotechnol.* **1995**, *13*, 142–150. [CrossRef]
6. Silverman, T.; Naffakh, M.; Marco, C.; Ellis, G. Morphology and thermal properties of biodegradable poly (hydroxybutyrate-co-hydroxyvalerate)/tungsten disulphide inorganic nanotube nanocomposites. *Mater. Chem. Phys.* **2016**, *170*, 145–153. [CrossRef]
7. Avella, M.; Martuscelli, E.; Raimo, M. Review Properties of blends and composites based on poly(3-hydroxy) butyrate (PHB) and poly (3-hydroxybutyrate-hydroxyvalerate)(PHBV) copolymers. *J. Mater. Sci.* **2000**, *35*, 523–545. [CrossRef]
8. Lagaron, J.M. *Multifunctional and Nanoreinforced Polymers for Food Packaging*, 1st ed.; Woodhead Publishing Elsevier Science: New York, NY, USA, 2011.
9. Muller, K.; Bugnicourt, E.; Latorre, M.; Jorda, M.; Sanz, Y.E.; Lagaron, J.M.; Miesbauer, O.; Bianchin, A.; Hankin, S.; Bolz, U.; et al. Review on the Processing and Properties of Polymer Nanocomposites and Nanocoatings and Their Applications in the Packaging, Automotive and Solar Energy Fields. *Nanomaterials* **2017**, *7*, 74. [CrossRef] [PubMed]
10. Fawaz, J.; Mittal, V. *Synthesis of Polymer Nanocomposites: Review of Various Techniques*, 1st ed.; John Wiley & Sons: Weinheim, Germany, 2014; pp. 1–26.

11. Carli, L.N.; Crespo, J.S.; Mauler, R.S. PHBV nanocomposites based on organomodified montmorillonite and halloysite: The effect of clay type on the morphology and thermal and mechanical properties. *Compos. Part A-Appl. Sci. Manuf.* **2011**, *42*, 1601–1608. [CrossRef]

12. Chen, G.X.; Hao, G.J.; Guo, T.Y.; Song, M.D.; Zhang, B.H. Structure and mechanical properties of poly(3-hydroxybutyrate-co-3-hydroxyvalerate) (PHBV)/clay nanocomposites. *J. Mater. Sci. Lett.* **2002**, *21*, 1587–1589. [CrossRef]

13. Ten, E.; Turtle, J.; Bahr, D.; Jiang, L.; Wolcott, M. Thermal and mechanical properties of poly(3-hydroxybutyrate-co-3-hydroxyvalerate)/cellulose nanowhiskers composites. *Polymer* **2010**, *51*, 2652–2660. [CrossRef]

14. Mook, C.W.; Wan, K.T.; Ok, P.O.; Keun, C.Y.; Woo, L.J. Preparation and characterization of poly(hydroxybutyrate-co-hydroxyvalerate)–organoclay nanocomposites. *J. Appl. Polym. Sci.* **2003**, *90*, 525–529. [CrossRef]

15. Öner, M.; İlhan, B. Fabrication of poly(3-hydroxybutyrate-co-3-hydroxyvalerate) biocomposites with reinforcement by hydroxyapatite using extrusion processing. *Mater. Sci. Eng. C* **2016**, *65*, 19–26. [CrossRef] [PubMed]

16. Xiang, H.X.; Chen, S.H.; Cheng, Y.H.; Zhou, Z.; Zhu, M.F. Structural characteristics and enhanced mechanical and thermal properties of full biodegradable tea polyphenol/poly(3-hydroxybutyrate-co-3-hydroxyvalerate) composite films. *eXPRESS Polym. Lett.* **2013**, *7*, 778–786. [CrossRef]

17. Luo, S.; Netravali, A.N. Mechanical and thermal properties of environment-friendly "green" composites made from pineapple leaf fibers and poly(hydroxybutyrate-co-valerate) resin. *Poly. Compos.* **1999**, *20*, 367–378. [CrossRef]

18. Du, J.; Zhao, G.; Pan; Zhuang, L.; Li, D.; Zhang, R. Crystallization and mechanical properties of reinforced PHBV composites using melt compounding: Effect of CNCs and CNFs. *Carbohydr. Polym.* **2017**, *168*, 255–262.

19. Qiang, T.; Wang, J.; Wolcott, M.P. Facile Fabrication of 100% Bio-Based and Degradable Ternary Cellulose/PHBV/PLA Composites. *Materials* **2018**, *11*, 330. [CrossRef] [PubMed]

20. Zhang, H.; Yu, H.Y.; Wang, C.; Yao, J. Effect of silver contents in cellulose nanocrystal/silver nanohybrids on PHBV crystallization and property improvements. *Carbohydr. Polym.* **2017**, *173*, 7–16. [CrossRef] [PubMed]

21. Wang, B.J.; Zhang, Y.J.; Zhang, J.Q.; Gou, Q.T.; Wang, Z.B.; Chen, P.; Gu, Q. Crystallization behavior, thermal and mechanical properties of PHBV/graphene nanosheet composites. *Chin. J. Polym. Sci.* **2013**, *31*, 670–678. [CrossRef]

22. Wang, D.; Chen, G.; Li, C.; Cheng, M.; Yang, W.; Wu, S.; Chen, P. Thermally induced graphene rotation on hexagonal boron nitride. *Phys. Rev. Lett.* **2016**, *116*, 126101–126105. [CrossRef] [PubMed]

23. Li, X.; Hao, X.; Zhao, M.; Wu, Y.; Yang, J.; Tian, Y.; Qian, G. Exfoliation of hexagonal boron nitride by molten hydroxides. *Adv. Mater.* **2013**, *25*, 2200–2204. [CrossRef] [PubMed]

24. Raman, C.; Meneghetti, P. Boron nitride finds new applications in thermoplastic compounds. *Plast. Addit. Comp.* **2008**, *10*, 26–31. [CrossRef]

25. Zhi, C.; Bando, Y.; Tang, C.; Kuwahara, H.; Golberg, D. Large-scale fabrication of boron nitride nanosheets and their utilization in polymeric composites with improved thermal and mechanical properties. *Adv. Mater.* **2009**, *21*, 2889–2893. [CrossRef]

26. Wattanakul, K.; Manuspiya, H.; Yanumet, N. Thermal conductivity and mechanical properties of BN-filled epoxy composite: Effects of filler content, mixing conditions and BN agglomerate size. *J. Compos. Mater.* **2011**, *45*, 1967–1980. [CrossRef]

27. Pradhan, G.C.; Behera, L.; Swain, S.K. Effects of boron nitride nanopowder on thermal, chemical and gas barrier properties of starch. *Chin. J. Polym. Sci.* **2014**, *32*, 1311–1318. [CrossRef]

28. Cheewawuttipong, W.; Fuoka, D.; Tanoue, S.; Uematsu, H.; Iemoto, Y. Thermal and mechanical properties of polypropylene/boron nitride composites. *Energy Procedia* **2013**, *34*, 808–817. [CrossRef]

29. Sun, G.; Bi, J.; Wang, W.; Zhang, J. Microstructure and mechanical properties of boron nitride nanosheets-reinforced fused silica composites. *J. Eur. Ceram. Soc.* **2017**, *37*, 3195–3202. [CrossRef]

30. Kisku, S.K.; Swain, S.K. Synthesis and characterization of chitosan/boron nitride composites. *J. Am. Ceram. Soc.* **2012**, *95*, 2753–2757. [CrossRef]

31. Zhou, W.; Zuo, J.; Zhang, X.; Zhou, A. Thermal, electrical and mechanical properties of hexagonal boron nitride-reinforced epoxy composites. *J. Compos. Mater.* **2014**, *48*, 2517–2526. [CrossRef]

32. Li, S.; Yang, T.; Zou, H.; Liang, M.; Chen, Y. Enhancement in thermal conductivity and mechanical properties via large-scale fabrication of boron nitride nanosheet. *High Perform. Polym.* **2016**, *29*, 315–327. [CrossRef]

33. Biscarat, J.; Bechelany, M.; Pochat-Bohatier, C.; Miele, P. Graphene-like BN/gelatin nanobiocomposites forgas barrier applications. *Nanoscale* **2015**, *7*, 613–618. [CrossRef] [PubMed]

34. Öner, M.; Çöl, A.A.; Pochat-Bohatier, C.; Bechelany, M. Effect of incorporation of boron nitride nanoparticles on the oxygen barrier and thermalproperties of poly(3-hydroxybutyrate-cohydroxyvalerate). *RSC Adv.* **2016**, *6*, 90973–90981. [CrossRef]

35. Öner, M.; Keskin, G.; Kızıl, G.; Pochat-Bohatier, C.; Bechelany, M. Development of poly(3-hydroxybutyrate-co-3-hydroxyvalerate)/boron nitride bionanocomposites with enhanced barrier properties. *Polym. Compos.* in press. [CrossRef]

36. Yu, H.; Sun, B.; Zhang, D.; Chen, G.; Yang, X.; Yao, J. Reinforcement of biodegradable poly(3-hydroxybutyrate-co-3-hydroxyvalerate) with cellulose nanocrystal/silver nanohybrids as bifunctional nanofillers. *J. Mater. Chem. B* **2014**, *2*, 8479–8489. [CrossRef]

37. Huang, W.; Wang, Y.; Ren, L.; Du, C.; Shi, X. A novel PHBV/HA microsphere releasing system loaded with alendronate. *Mater. Sci. Eng. C* **2009**, *29*, 2221–2225. [CrossRef]

38. Ambrosio-Martín, J.; Gorrasi, G.; Lopez-Rubio, A.; Fabra, M.J.; Mas, L.C.; López-Manchado, M.A.; Lagaron, J.M. On the use of ball milling to develop poly(3-hydroxybutyrate-co-3-hydroxyvalerate)-graphene nanocomposites (II)—Mechanical, barrier, and electrical properties. *J. Appl. Polym. Sci.* **2015**, *132*, 42217–42225. [CrossRef]

39. Yu, H.; Yan, C.; Yao, J. Fully biodegradable food packaging materials based on functionalized cellulose nanocrystals/poly(3-hydroxybutyrate-co-3-hydroxyvalerate) nanocomposites. *RSC Adv.* **2014**, *4*, 59792–59802. [CrossRef]

40. Sato, H.; Murakami, R.; Padermshoke, A.; Hirose, F.; Senda, K.; Noda, I.; Ozaki, Y. Infrared Spectroscopy Studies of CH–O Hydrogen Bondings and Thermal Behavior of Biodegradable Poly(hydroxyalkanoate). *Macromolecules* **2004**, *37*, 7203–7213. [CrossRef]

41. Suthar, V.; Pratap, A.; Raval, H. Studies on poly(hydroxy alkanoates)/(ethylcellulose) blends. *Bull. Mater. Sci.* **2000**, *23*, 215–219. [CrossRef]

42. Singh, S.; Mohanty, A.K.; Sugie, T.; Takai, Y.; Hamada, H. Renewable resource based biocomposites from natural fiber and polyhydroxybutyrate-co-valerate (PHBV) bioplastic. *Compos. Part A-Appl. Sci. Manuf.* **2008**, *39*, 875–888. [CrossRef]

43. Yu, L.; Gao, B.; Chen, Z.; Sun, C.; Cui, D.; Wang, C.; Wang, Q.; Jiang, M. In Situ FTIR Investigation on Phase Transformations in BN Nanoparticles. *Chin. Sci. Bull.* **2005**, *50*, 2827–2831.

44. Shi, L.; Gu, Y.; Chen, L.; Qian, Y.; Yang, Z.; Ma, J. Synthesis and Morphology Control of Nanocrystalline Boron Nitride. *J. Solid State Chem.* **2004**, *177*, 721–724. [CrossRef]

45. Soliveri, G.; Pifferi, V.; Annunziata, R.; Rimoldi, L.; Aina, V.; Cerrato, G.; Meroni, D. Alkylsilane–SiO$_2$ Hybrids. A Concerted Picture of Temperature Effects in Vapor Phase Functionalization. *J. Phys. Chem. C* **2015**, *119*, 15390–15400. [CrossRef]

46. Paul, B.; Martens, W.N.; Frost, R.L. Organosilane grafted acid-activated beidellite clay for the removal of non-ionic alachlor and anionic imazaquin. *Appl. Surf. Sci.* **2011**, *257*, 5552–5558. [CrossRef]

47. Li, J.; Lai, M.F.; Liu, J.J. Effect of poly (propylene carbonate) on the crystallization and melting behavior of poly(hydroxybutyrate-co-hydroxyvalerate). *J. Appl. Polym. Sci.* **2004**, *92*, 2514–2521. [CrossRef]

48. Camargo, F.A.; Innocentini-Mei, L.H.; Lemes, A.P.; Moraes, S.G.; Durán, N. Processing and characterization of composites of poly (3-hydroxybutyrate-co-hydroxyvalerate) and lignin from sugar cane bagasse. *J. Compos. Mater.* **2012**, *46*, 417–425. [CrossRef]

49. Bhardwaj, R.; Mohanty, A.K.; Drzal, L.T.; Pourboghrat, F.; Misra, M. Renewable resource-based green composites from recycled cellulose fiber and poly(3-hydroxybutyrate-co-3-hydroxyvalerate) bioplastic. *Biomacromolecules* **2006**, *7*, 2044–2051. [CrossRef] [PubMed]

50. Lai, M.; Li, J.; Yang, J.; Liu, J.; Tong, X.; Cheng, H. The morphology and thermal properties of multi-walled carbon nanotube and poly(hydroxybutyrate-co hydroxyvalerate) composite. *Polym. Int.* **2004**, *53*, 1479–1484. [CrossRef]

51. Avella, M.; La, R.G.; Martuscelli, E.; Raimo, M. Poly(3-hydroxybutyrate-co-3-hydroxyvalerate) and wheat straw fibre composites: Thermal, mechanical properties and biodegradation behavior. *J. Mater. Sci.* **2000**, *35*, 829–836. [CrossRef]

52. Hsu, S.F.; Wu, T.M.; Liao, C.S. Nonisothermal crystallization behavior and crystalline structure of poly(3-hydroxybutyrate)/layered double hydroxide nanocomposites. *J. Polym. Sci. Part B: Polym. Phys.* **2007**, *45*, 995–1002. [CrossRef]

53. Cretois, R.; Follain, N.; Dargent, E.; Soulestin, J.; Bourbigot, S.; Marais, S.; Lebrun, L. Microstructure and barrier properties of PHBV/organoclays bionanocomposites. *J. Membr. Sci.* **2014**, *467*, 56–66. [CrossRef]

54. Sanchez-Garcia, M.D.; Gimenez, E.; Lagaron, J.M. Novel PET nanocomposites of interest in food packaging applications and comparative barrier performance with biopolyester nanocomposites. *J. Plast. Film Sheet.* **2007**, *23*, 133–148. [CrossRef]

55. Hassaini, L.; Kaci, M.; Touati, N.; Pillin, I.; Kervoelen, A.; Bruzaud, S. Valorization of olive husk flour as a filler for biocomposites based on poly (3-hydroxybutyrate-co-3-hydroxyvalerate): Effects of silane treatment. *Polym. Test.* **2017**, *59*, 430–440. [CrossRef]

56. Pokharel, P.; Bae, H.; Lim, J.; Yong, L.K.; Choi, S. Effects of titanate treatment on morphology and mechanical properties of graphene nanoplatelets/high density polyethylene nanocomposites. *J. Appl. Polym. Sci.* **2015**, *132*, 42073–42085. [CrossRef]

57. Lu, Y.L.; Zhang, G.B.; Feng, M.; Zhang, Y.; Yang, M.S.; Shen, D.T. Hydrogen bonding in polyamide 66/clay nanocomposite. *J. Polym. Sci. Part B-Polym. Phys.* **2003**, *41*, 2313–2321. [CrossRef]

58. Evans, J.R.G.; Chen, B. Poly(epsilon-caprolactone)-clay nanocomposites: Structure and mechanical properties. *Macromolecules* **2006**, *39*, 747–754.

59. Yu, W.; Lan, C.H.; Wang, S.J.; Fang, P.F.; Sun, Y.M. Influence of zinc oxide nanoparticles on the crystallization behavior of electrospun poly(3-hydroxybutyrate-co-3-hydroxyvalerate) nanofibers. *Polymer* **2010**, *51*, 2403–2409. [CrossRef]

60. Bhattacharya, M. Polymer Nanocomposites—A comparison between Carbon Nanotubes, Graphene and Clay as Nanofillers. *Materials* **2016**, *9*, 262. [CrossRef] [PubMed]

61. Cadek, M.; Coleman, J.N.; Ryan, K.P.; Nicolosi, V.; Bister, G.; Fonseca, A.; Nagy, J.B.; Szostak, K.; Beguin, F.; Blau, W.J. Reinforcement of polymers with carbon nanotubes: The role of surface area. *Nano Lett.* **2004**, *4*, 353–356. [CrossRef]

62. Carotenuto, G.; De Nicola, S.; Palomba, M.; Pullini, D.; Horsewell, A.; Hansen, T.W.; Nicolais, L. Mechanical properties of low-density polyethylene filled by graphite nanoplatelets. *Nanotechnology* **2012**, *23*, 48570–48713. [CrossRef] [PubMed]

63. Xiang, H.; Li, L.; Wang, S.; Wang, R.; Cheng, Y.; Zhou, Z.; Zhu, M. Natural polyphenol tannic acid reinforced poly(3-hydroxybutyrate-co-3-hydroxyvalerate) composite films with enhanced tensile strength and fracture toughness. *Polym. Compos.* **2015**, *36*, 2303–2308. [CrossRef]

64. Dong, Y.; Bhattacharyya, D. A simple micromechanical approach to predict mechanical behaviour of polypropylene/organoclay nanocomposites based on representative volume element (RVE). *Comput. Mater. Sci.* **2010**, *49*, 1–8. [CrossRef]

65. Colemn, J.N.; Khan, U.; Blau, W.J.; Gun'ko, Y.K. Small but strong: A review of the mechanical properties of carbon nanotube–polymer composites. *Carbon* **2006**, *44*, 1624–1652. [CrossRef]

nanomaterials

MDPI

Article

Hexagonal Boron Nitride Functionalized with Au Nanoparticles—Properties and Potential Biological Applications

Magdalena Jedrzejczak-Silicka [1,*], Martyna Trukawka [2], Mateusz Dudziak [2], Katarzyna Piotrowska [3] and Ewa Mijowska [2,*]

[1] Laboratory of Cytogenetics, West Pomeranian University of Technology, Szczecin, Klemensa Janickiego 29, 71-270 Szczecin, Poland
[2] Nanomaterials Physicochemistry Department, West Pomeranian University of Technology, Szczecin, Piastow Avenue 45, 70-311 Szczecin, Poland; martyna.brylak@zut.edu.pl (M.T.); mateusz.dudziak@zut.edu.pl (M.D.)
[3] Department of Physiology, Pomeranian Medical University in Szczecin, Powstancow Wlkp. 72, 70-111 Szczecin, Poland; piot.kata@gmail.com
* Correspondence: mjedrzejczak@zut.edu.pl (M.J.-S.); emijowska@zut.edu.pl (E.M.); Tel.: +48-914-496-804 (M.J.-S.); +48-914-494-742 (E.M.)

Received: 9 July 2018; Accepted: 4 August 2018; Published: 9 August 2018

Abstract: Hexagonal boron nitride is often referred to as white graphene. This is a 2D layered material, with a structure similar to graphene. It has gained many applications in cosmetics, dental cements, ceramics etc. Hexagonal boron nitride is also used in medicine, as a drug carrier similar as graphene or graphene oxide. Here we report that this material can be exfoliated in two steps: chemical treatment (via modified Hummers method) followed by the sonication treatment. Afterwards, the surface of the obtained material can be efficiently functionalized with gold nanoparticles. The mitochondrial activity was not affected in L929 and MCF-7 cell line cultures during 24-h incubation, whereas longer incubation (for 48, and 72 h) with this nanocomposite affected the cellular metabolism. Lysosome functionality, analyzed using the NR uptake assay, was also reduced in both cell lines. Interestingly, the rate of MCF-7 cell proliferation was reduced when exposed to h-BN loaded with gold nanoparticles. It is believed that h-BN nanocomposite with gold nanoparticles is an attractive material for cancer drug delivery and photodynamic therapy in cancer killing.

Keywords: boron nitride; nanocomposite; functionalization; gold nanoparticles; cytobiocompatibility

1. Introduction

Recently, graphene, related two-dimensional crystals and hybrid particles have been intensively studied in the context of technological and scientific evolution. Unique properties of graphene allow for using it for various purposes in many fields (e.g., biomedical applications, energy storage, electronic devices, biosensors, spintronics or photonics) [1]. Hexagonal boron nitride (h-BN) is one of the 2D layered materials with specific properties [1]. BNs were considered only as synthetics, but recently they have also been discovered in the natural environment (Qingsongite (IMA2013-30)—natural analogue of cubic boron nitride) [2,3]. Hexagonal boron nitride is an analogue of graphite. In its structure, alternating B and N atoms substitute C atoms [4]. Boron and nitrogen atoms are linked with each other via strong B-N covalent bonds to form interlocking hexagonal rings [2,4]. The 2D layers of h-BN are held together by weak van der Waals forces [2,5,6]. The B-N bond length is 1.466 Å, whereas the interlayer space is 3.331 Å [5]. In comparison to graphene, where C-C are covalently bonded, h-BN covalent B-N bonds are partially ionic; this is due to B atoms, which in every consecutive BN layer are positioned exactly above or below N atoms in the adjacent layers. Such structural characteristics implies the polarity of B-N bonds [2,4]. Boron nitride can also exhibit a hollow spherical morphology [4,7], diamond-like cubic form (c-BN), with boron and nitrogen atoms forming a tetrahedral bond network, as well as wurtzite BN (w-BN). The cubic structure with alternating boron and nitrogen atoms is similar to that of diamond [5]. Its flakes may be mono- or several-layer-thick. The formation of multilayer stabilizes the whole structure. h-BN systems (e.g., nanotubes, flakes) show chemical and thermal stability, but at the same time they are equally thermally conductive and mechanically robust. h-BN is an electrical insulator with a band gap of ~5–6 eV [4].

h-BN is used in different fields due to its interesting physical and chemical properties, e.g., in electronics as an insulator, substrate for semi-conductors, coating for refractory molds, in ceramics, resins, plastics (to obtain self-lubricating properties) [5,7,8]. BN nanosheets were found useful in polymeric film reinforcement, e.g., the elastic modulus of polymethyl methacrylate (PMMA) film was increased when BN nanosheets were incorporated into the polymer [4]. Hexagonal boron nitride is also widely used in the production of coatings and paintings for high temperature applications. Boron nitride is also a popular inorganic compound in cosmetic industry used as a slip modifier [5,7]. Lately, hexagonal boron nitride has been found to be excellent substrate material [9,10]. The h-BN platforms are used in creation of a new generation of few-atomic-layer vdW (van der Waals) heterostructures. Most 2D heterostructures (vertical heterostructures) are synthesized by stacking of individual layers of different materials [10,11]. Unlimited van der Waals interplanar interactions in layered materials provide possibility to integrate with an array any layered 2D material (such as graphene, hexagonal boron nitride, transition-metal dichalcogenides) of different dimensionality (e.g., combinations of 2D + nD materials, where n = 0, 1 and 3) [12]. Zhao and co-workers (2018) presented controlled electrochemical intercalation of graphene/h-BN vdW heterostructures, where Li was electrochemical intercalated into graphene encapsulated between h-BN layers resulting in higher carrier density [9,13]. Using of exfoliated h-BN crystals [14] as a substrate, crystalline high-quality rubene have been template. Obtained heterostructure enabling creation of organic FETs (OFETs) with carrier motilities exceeding 10 cm^2 V^{-1} s^{-1} [12,14,15]. By comparison, graphene on ultra-flat boron nitride (BN) has shown intrinsic mobility approaching 500,000 cm^2 V^{-1} s^{-1} [16]. Those structures display unique properties, diverse functionality, great potential and can response to the need of electronic and electrochemical industry [13]. Due to their ultrasensitivity, 2D heterostructures present a broad range of applications, e.g., photovoltaic [12], field effect/tunnelling transistors, optronics [17], photodetectors, light-emitting [18], electronic [19], thermoelectric and memory [15,19] devices and bio-sensing [20].

Boron nitride exhibits also hydrophobicity in aqueous environment [21]. Therefore, it seems to be suitable for biomedical applications after specific functionalization process. The problem of limited BNs dispersion is one of the most challenging approaches [22]. Several cytotoxicity studies based on boron nitride nanotubes (BNNT), hollow boron nitride nanospheres, h-BN nanosheets confirmed its low cytotoxicity and suggested that BN can be used as a novel drug delivery system. In contrast, other studies have showed that BNNT had cytotoxic effect and affected relative cell viability even at low concentrations [23–26].

Although boron nitrides have unique and applicable properties, the number of BN-related publications is significantly smaller in comparison to the widely studied C systems [3]. Thus the aim of the study was to evaluate exfoliated hexagonal boron nitride functionalized with Au nanoparticles, for potential biomedical applications.

2. Materials and Methods

2.1. Materials

Hexagonal boron nitride, gold(III) chloride trihydrate, phosphate buffered saline, polyethylene glycol, Pluronic F-127, dehydrate trisodium citrate were purchased from Sigma-Aldrich (St. Louis, MO, USA). Hydrogen peroxide solution, sulfuric acid, potassium permanganate and 1-methyl-2-pyrrolidinone were obtained from Chempur (Piekary Slaskie, Poland).

2.2. Methods

2.2.1. Exfoliation of h-BN

Chemical exfoliation of h-BN was carried out by a modified Hummer's method, similar to graphite exfoliation. 750 mg of h-BN was mixed with 3.0 g of potassium permanganate in a three-neck flask. The whole system was installed under the reflux. Next, 60 mL of 96% sulfuric acid was slowly added. The mixture was heated at 40 °C for 6 h. Subsequently, the system was cooled down. The flask with the mixture was inserted into the ice bath. Then 200 mL of hydrogen peroxide solution was added. After this process, the mixture was purified. Purification was carried out via multiple water washing and centrifugation at 10,000 rpm for 15 min until the pH reached 7.

Chemically exfoliated h-BN was additionally delaminated mechanically. Mechanical exfoliation was performed using a tip sonicator. Chemically exfoliated h-BN was added into 1-methyl-2-pyrrolidinone (NMP) in a volume ratio of 0.5% and sonicated (600 W 25%) for 30 min with a pulse mode of 5s on/5s off. After sonication, the mixture was left to evaporate the solvent.

2.2.2. Hexagonal Boron Nitride Au Functionalization

Exfoliated h-BN was functionalized with gold nanoparticles. 100 mL of distilled water was mixed with 6 mg of h-BN. The mixture was heated at 100 °C under the reflux. Next 4 mL of gold(III) chloride trihydrate was added at a concentration of 2 mg mL^{-1}. After a few minutes, 40 mg of trisodium citrate was added to the boiling content. The whole system was heated for 1 h at 100 °C. After 1 h, the mixture was cooled down for the purification. The purification was performed by multiple washing with distilled water and centrifugation at 8000 rpm for 10 min until the pH reached 7.

2.3. Characterization of Synthesized Nanomaterial

The samples were examined using transmission electron microscopy (TEM, FEI Tecnai F30, Frequency Electronics Inc., Thermo Fisher Scientific, Waltham, MA, USA). The phase composition of samples was characterized by X-ray diffraction (XRD) analysis (X'Pert PRO Philips diffractometer, Almelo, The Netherlands) using a CoK$_\alpha$ radiation. UV-Vis absorption spectra of h-BN, h-BN nanocomposite (h-BN_AuNP) and gold nanoparticles were recorded with a Helios alpha UV-Vis Spectrometer (Thermo Fisher Scientific, Waltham, MA, USA). Fourier transform infrared (FT-IR) absorption spectra were measured on a Nicolet 6700 FT-IR spectrometer (Thermo Nicolet Corp., Madison, WI, USA). Raman spectra were measured with a Renishaw in via Raman microscope at 785 nm. The surface and thickness of the flakes were measured by atomic force microscopy (AFM, Nanoscope V Multimode 8, Bruker, Mannheim, Germany). Zeta potential was measured by Zeta Sizer (ZS Nano, Malvern Panalytical, Malvern, UK).

2.4. Dispersion Stability of Au Functionalized h-BN

Before cytocompatibility analyses, the dispersion stability of hexagonal boron nitride was examined in phosphate buffered saline (PBS) with dispersant Pluronic F-127. Concentration of PBS-Pluronic F-127 was 1 mg mL^{-1}. Subsequently, different amounts of nanomaterial (12.5 μg mL^{-1}, 25 μg mL^{-1}, 50 μg mL^{-1} and 100 μg mL^{-1}, respectively) was diluted with PBS-polymer and sonicated to obtain homogeneous solution of the following concentration. The UV-Vis monitoring (Thermo Scientific GENESYS 10S, Thermo Fisher Scientific, Waltham, MA, USA) at 350 nm was evaluated to determine the dispersions stability after 1, 3, 5, 20, 22, 24, 44, 48 and 51 h.

2.5. Cell Lines and Cell Culture Conditions

Two adherent cell lines—murine L929 fibroblast (ATCC® CCL-1™, American Type Culture Collection, Manassas, VA, USA) and MCF-7 human breast adenocarcinoma (ATCC® HTB-22™,

American Type Culture Collection, Manassas, VA, USA)—were chosen for cytocompatibility studies of the h-BN_AuNP nanocomposite.

For morphology analyses (phase contrast and holographic microscopy), cells of each line were seeded into T25 flasks (Sarstedt, Nümbrecht, Germany) and maintained in standard cell culture conditions at 37 °C, 5% CO_2, 95% humidity. Complete Dulbecco's Modified Eagle Medium (DMEM) culture medium supplemented with 10% heat inactivated fetal bovine serum (FBS) (PAA Laboratories GmbH, Pasching, Austria), 2 mM L-glutamine, 50 IU mL^{-1} penicillin and 50 µg mL^{-1} streptomycin (Sigma-Aldrich, St. Louis, MO, USA) was used in the study.

For cytocompatibility analysis, the cells were seeded into 96-well plates (Corning Inc., New York, NY, USA) and cultured for 24, 48 and 72 h in standard conditions mentioned above. All cell cultures were monitored with a Nikon TS-100 microscope (Nikon, Melville, NY, USA).

2.6. Experimental Treatment

For experimental treatment, the h-BN_AuNP nanocomposite was prepared in a 100 µg mL^{-1} Pluronic F-127 solution in PBS to a stock concentration of 1 mg mL^{-1}. Twenty-four hours after cell seeding, 8 different final concentrations (3.125, 6.25, 10.0, 12.5, 25.0, 50.0, 100.0, 200.0 µg mL^{-1}) of h-BN_AuNP nanocomposite were prepared in DMEM medium and added to the cell cultures. Additionally, the Pluronic F-127 vehicle control (only DMEM medium with dispersant) was prepared for the reference dispersion as well as the standard control culture (cells cultured in standard DMEM medium in the absence of the h-BN_AuNP nanocomposite and Pluronic F-127). Cell lines were incubated with the h-BN_AuNP nanocomposite for 24, 48 and 72 h.

2.7. Microscopic Analyses

Firstly, the morphology of L929 and MCF-7 cell lines exposed to the h-BN_AuNP nanocomposite at different concentrations and control samples was analyzed using a Nikon TS-100 phase contrast inverted microscope (Nikon, Melville, NY, USA) at 400× magnification.

Secondly, holographic microscopy label-free images were obtained using the HM4 HoloMonitor™ (Phase Holographic Imaging, Lund, Sweden). For presented analyses, the HoloMonitor™ connected to the computer was placed inside a CO_2 incubator (Memmert GmbH, Germany) to capture time-lapse image sequences during the cell treatment. Similarly to the phase contrast microscopy, the images using the HoloMonitor M4 were taken for different concentrations of the h-BN_AuNP nanocomposite and the control sample for both cell lines. The time-laps image sequence was recorded every 1 min for 24–72 h (24 h after cell seeding). The doubling time (DT) was established basing on holographic observations, as a measure of cell growth for each cell line using the following Formula (1):

$$\text{Doubling Time} = \frac{\text{duration} \times \log(2)}{\log(\text{Final concentration}) - \log(\text{Initial concentration})} \quad (1)$$

2.8. Cell Counting Kit-8 Analysis

The relative mitochondrial activity of L929 and MCF-7 cell lines after 24, 48 and 72-h incubation with the h-BN_AuNP nanocomposite was tested using the CCK-8 Cell Counting Kit-8 (Sigma-Aldrich, St. Louis, MO, USA). The CCK-8 assay is based on the conversion of tetrazolium salt into the colored formazan by the living cells. The amount of the reduction product is proportional to the number of metabolically active cells. The CCK-8 solution (10 µL) was added to each well and incubated for 240 min at 37 °C in an incubator. After the incubation period, the absorbance was recorded at 450 nm, according to the manufacturer's instructions using a Sunrise Absorbance Reader (Tecan, Männedorf, Switzerland). All experiments were conducted in triplicate.

The effect of the nanocomposite on cellular metabolic activity was calculated using the following Formula (2):

$$\text{Relative viability from the CCK}-8 \text{ assay}(\%) = \left(\frac{\text{sample abs}_{450\text{--}650\text{ nm}}}{\text{positive control abs}_{450\text{--}650\text{ nm}}}\right) \times 100 \qquad (2)$$

2.9. Lactate Dehydrogenase Leaking Assay

Cytocompatibility of the h-BN_AuNP nanocomposite was evaluated using the LDH CytoTox 96® Non-Radioactive Cytotoxicity Assay (Promega, Madison, WI, USA). The LDH CytoTox 96® Non-Radioactive Cytotoxicity Assay measures lactate dehydrogenase released due to cellular membrane damage. The amount of formazan converted from tetrazolium salt is proportional to the number of lysed cells. The LDH assay was performed according to the manufacturer's instructions (Promega, Madison, WI, USA) and the absorbance was measured at 490 nm using a microplate spectrophotometer (Absorbance Reader, Tecan, Männedorf, Switzerland). The interaction between the solution with the nanocomposite in the cell culture medium and LDH assay components was tested in the absence of cells. The percentage of LDH released after 24, 48 and 72-h exposure was calculated using the Formula (3):

$$\%\text{LDH released} = \frac{A490 \text{ nm of treated and untreated cells} - A490 \text{ nm of control}}{A490 \text{ nm of maximum of untreated cells} - A490 \text{ nm of control}} \times 100 \qquad (3)$$

where *A* is absorbance.

2.10. Neutral Red Uptake Assay

Similarly, the neutral red uptake assay (NRU) (In vitro Toxicology Assay Kit, Neutral Red based, Sigma-Aldrich, St. Louis, MO, USA) was performed 24–72 h after the L929 and MCF-7 cell exposure to the tested nanocomposite. The neutral red uptake assay is based on the viable cell ability to store the neutral red dye in acidic organelles by the active transport (e.g., in lysosomes). Fresh DMEM medium containing 10% of neutral red was added to the cultures and incubated at 37 °C, 5% CO_2 and 95% relative humidity for 3 h. After the incubation, the cells were washed with DPBS and the Solubilization Solution was added to release the incorporated dye from the cells. The culture plates were allowed to rest for 10 min at room temperature, then gentle stirred and the absorbance at 540 nm was measured using a Tecan Sunrise microplate reader (Tecan, Männedorf, Switzerland). The effect of the nanocomposite on cell viability was calculated using the following Equation (4):

$$\text{Neutral red uptake assay}(\%) = \left(\frac{\text{sample abs}_{540\text{--}690\text{ nm}}}{\text{positive control abs}_{540\text{--}690\text{ nm}}}\right) \times 100 \qquad (4)$$

2.11. Cellular Uptake and Confocal Microscope Imaging

For the cellular uptake analysis, cells were plated in chamber slides (Lab-Tek Chamber, 4-wells, Thermo Scientific, Waltham, MA, USA) and cultured for 24 h. 24 h after the cell seeding, 50 µg mL^{-1} h-BN nanoflakes labeled with FITC (Sigma-Aldrich, St. Louis, MO, USA) were added to the culture medium and incubated for additional 24–72 h. Afterwards, the cells were washed three times with DPBS (PAN-Biotech GmbH, Aidenbach, Germany) and they were fixed with 4% paraformaldehyde solution (Sigma-Aldrich, St. Louis, MO, USA) for 10 min at RT. Next, for the cellular nuclei localization cells were stained with DAPI solution (5 µg mL^{-1}, Sigma-Aldrich, St. Louis, MO, USA) for 20 min at RT. The microphotographs were collected in FV1000 Confocal system with Olympus IX81 inverted microscope (Olympus, Hamburg, Germany) in two separated channels: for DAPI (405 nm diode), for FITC (488 nm laser). The microphotographs are showed as merged image.

2.12. Statistical Analysis

Data collected in CCK-8, LDH, and NRU assays are given as the mean values ± standard deviation (SD) and analyzed using ANOVA. The statistical analyses for CCK-8 and LDH assay results were performed using Levene's test of homogeneity. For data obtained from NRU assays Kruskal-Wallis analyses were conducted. The *p*-values < 0.05 were considered significant and are represented by different small letters (Figure 8A–F). Statistical analyses were performed using the *STATISTICA* 12.5 (StatSoft Inc., Tulsa, OK, USA) software.

3. Results

3.1. Characterization of h-BN and h-BN_AuNP

The morphology of h-BN was evaluated using transmission electron microscopy. The most important was to assess the shape and thickness of h-BN flakes before and after the exfoliation process. The commercial h-BN flakes were very thick (~300 nm) [27,28], which corresponded to 900 layers in the individual flake. A large number of flakes was aggregated and connected to each other. After combining chemical and mechanical exfoliation methods, thin flakes have been obtained (shown in Figure S1A, Supplementary materials). Atomic force microscopic analyses also confirmed this feature of h-BN. By using AFM, it is possible to calculate thickness of a single h-BN flake. Thickness of bulk h-BN is about 166 layers in individual flake. The morphology and thickness of exfoliated h-BN is shown in Figure 1A,B. Based on our calculations, the flake thickness after exfoliations was about 5 nm.

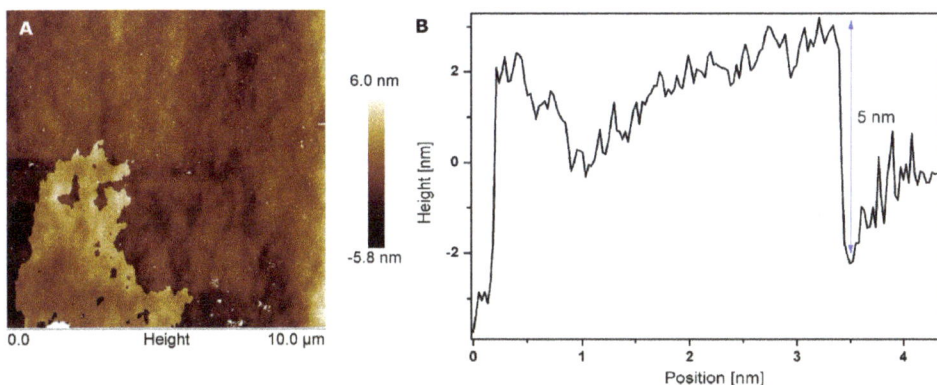

Figure 1. Atomic force microscope analysis for surface morphology (**A**) and flake thickness (**B**) of exfoliated h-BN.

In order to analyze the presence of the functional groups induced in the exfoliation process, FT-IR spectra were obtained for commercial h-BN and h-BN exfoliated by combined methods (chemical and physical exfoliation). Figure 2 shows the differences in spectra between bulk and exfoliated h-BN. No functional groups were found in commercial h-BN. O-H groups appeared after chemical exfoliation, with peaks near 2526 and 2330 cm^{-1}. The peaks at 1367 cm^{-1} and 810 cm^{-1} are derived from hexagonal boron nitride and represented B-N bonds.

To confirm the interactions of gold nanoparticles and h-BN, a Fourier Transform Infrared Spectroscopy (FTIR) study was conducted. The spectra of h-BN and h-BN_Au are compared in Figure S2 (Supplementary materials). There are three characteristic peaks in h-BN. The peak at 810 cm^{-1} corresponds to B-N vibrations. And the band at 940 cm^{-1} is assigned to B-N-O [29]. The peak at 1370 cm^{-1} is due to the stretching vibrations of B-N. The hydroxyl band was observed at 2525 and 3400 cm^{-1} for h-BN. After the Au functionalization these peaks disappeared. Interestingly, it is noteworthy that the sample of h-BN with Au shows a new band near 1100 cm^{-1}, which corresponds in particular to C–O stretching vibrations [30]. That is, when Au^{3+} ions are reduced with the hydroxyl groups of the boron nitride, the hydroxyl groups are oxidized to carbonyl group. It can be concluded from the above results that Au^{3+} ions can be reduced by hydroxyl groups of h-BN, resulting in the formation of gold nanoparticles on the surface [31]. It is also worth mentioning that the peak at 1370 cm^{-1} changed its shape and width, what can also confirm the interactions between h-BN and gold nanoparticles.

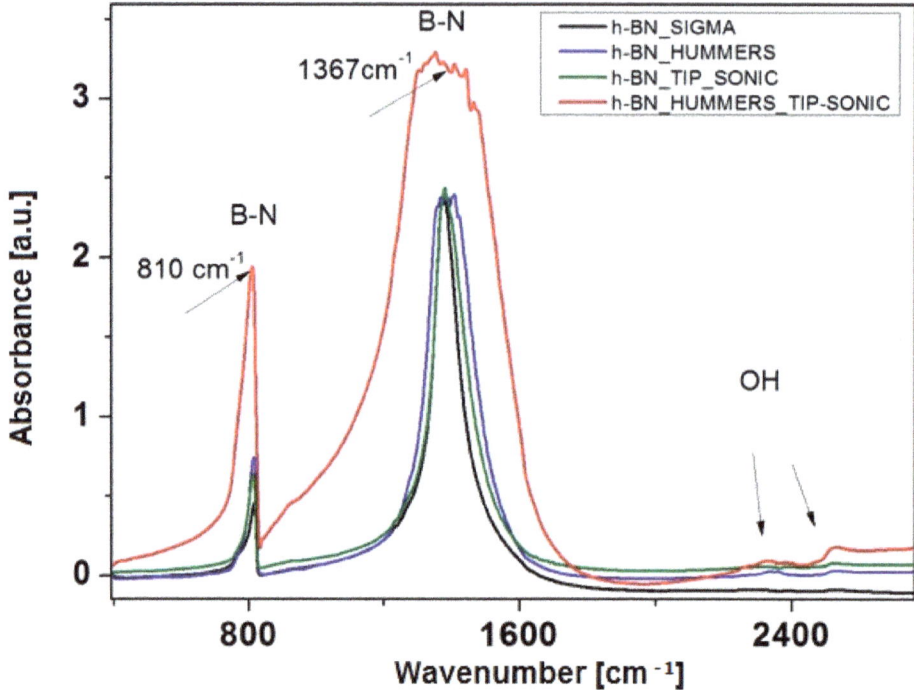

Figure 2. FT-IR spectra of bulk and exfoliated h-BN.

Morphology and thickness of h-BN_AuNP nanocomposite is shown in Figure 3. Transmission electron microscopy confirmed that gold nanoparticles were deposited on the h-BN surface. The size of gold particles ranged from 10 to 20 nm with the average value of 12 nm (about 30.5%; Figure 3D). There was not many agglomerates of gold nanoparticles. Most of nanoparticles were homogeneously distributed on h-BN platform. Thickness of the nanocomposite was similar to the exfoliated h-BN. The morphology of h-BN_AuNP is shown in Figure 3.

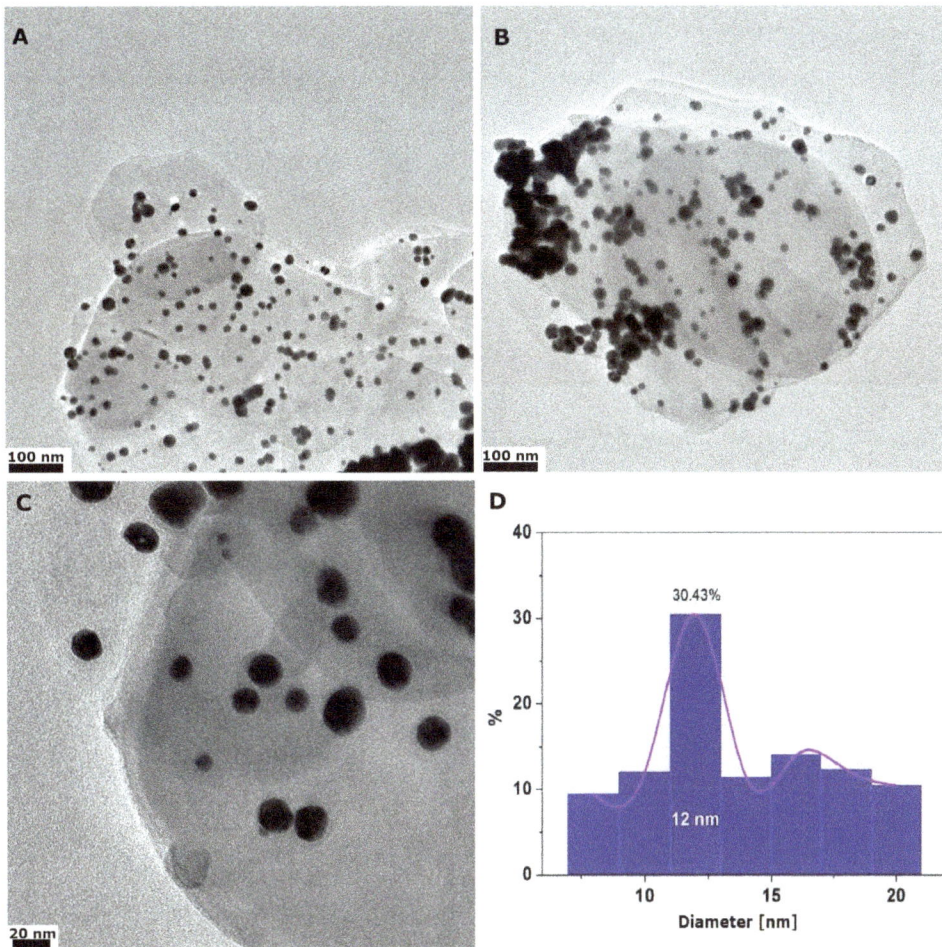

Figure 3. Transmission electron microscope images of h-BN_AuNP nanocomposite (**A–C**) and histogram of particle size distribution (**D**).

The XRD analysis and selected area electron diffraction pattern of the h-BN exfoliated and h-BN_AuNP nanocomposite are shown in Figure 4A. On h-BN XRD pattern there are four peaks at the Bragg angles of 26.52 2Θ; 41.3 2Θ; 54.8 2Θ; 76 2Θ, that correspond to (002), (100), (004) and (110) planes of h-BN, respectively. These diffraction peaks prove the hexagonal structure of boron nitride [32].

With regard to the h-BN_AuNP nanocomposites, Bragg angles of 38.3 2Θ; 44.3 2Θ; 64.5 2Θ and 77.6 2Θ were clearly observed, which correspond to the (111), (200), (220) and (311) planes of the AuNPs. There are also characteristic diffraction peaks coming from h-BN. These results confirmed that the AuNPs anchored on h-BN were well crystallized. The AuNPs/h-BN nanocomposite was successfully prepared [33]. The electron diffractions pattern (Figure 4B,C) confirms the results obtained from X-ray diffractometer.

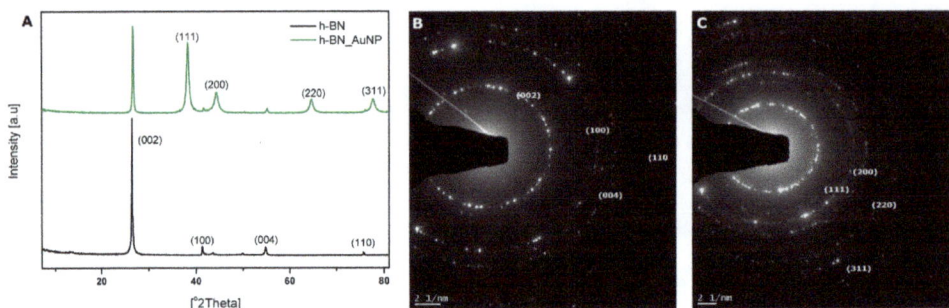

Figure 4. XRD patterns (**A**), electron diffraction patterns of h-BN (**B**) and h-BN_AuNP (**C**).

Raman spectra (Figure 5) were measured for 3 samples: exfoliated h-BN, h-BN_AuNP nanocomposite and gold nanoparticles. The most characteristic peak in h-BN at 1366 cm^{-1} is due to the E2g phonon mode and analogous to the G peak in graphene [4,10,34,35]. It can be seen in the spectra of h-BN and Au nanocomposite spectra. The peak is very intensive in h-BN spectrum. Its intensity is much smaller in the nanocomposite due to presence of gold nanoparticles—they have characteristic peaks in the region from 1300 to 1600 cm^{-1}. Similar observations were made by Draz et al. (2016) in their research [36]. However, the most prominent peak confirming the functionalization of the h-BN by gold nanoparticle is at about 2130 cm^{-1}. It is present in the spectrum of pristine gold nanoparticles and the nanocomposite with h-BN.

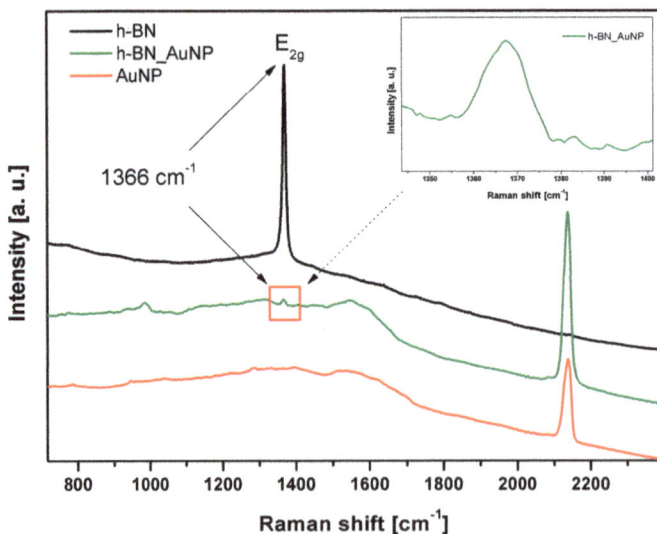

Figure 5. Raman spectra of h-BN, h-BN_AuNP nanocomposite and AuNP.

The stability of h-BN_AuNP nanocomposite is shown in Figure 6. A UV-Vis spectrophotometer was used for its evaluation. The solution was prepared by the dissolving one tablet of PBS in 200 mL of water. Next, the solution of Pluronic F-127 100 µg mL^{-1} was prepared. The h-BN_AuNP nanocomposite was added to the polymer solution in PBS to the final concentrations of 12.5, 25, 50 and 100 µg mL^{-1}, respectively. The dispersion stability was verified for 48 h. The best results were obtained at the concentration of 12.5 µg mL^{-1}. However, the dispersion was stable also at higher concentrations.

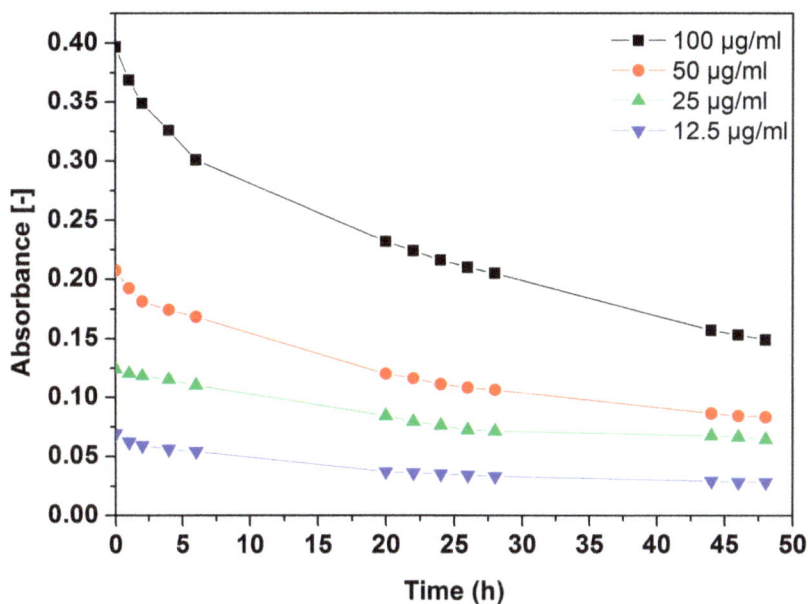

Figure 6. Dispersion stability of h-BN_AuNP with nanomaterial concentration range of 12.5–100 μg mL^{-1}.

In order to determine what the loading of gold in the nanocomposite, the UV-Vis spectrophotometry was used. The peak near 530 nm indicated that there were particles about 10 nm in size. Figure S3A (Supplementary materials) represents spectrum of pristine h-BN showing only one peak ~200 nm, while the h-BN_AuNP nanocomposite exhibited also a peak at 530 nm. The simple quantitative analysis allowed to conclude that there was ~16 wt% of the nanocomposite.

The UV-Vis studies were also performed to determine the stability of the interaction between hexagonal boron nitride and gold nanoparticles. At appropriate time intervals, the nanocomposite solution was tested on a UV-Vis spectrometer to determine the possible changes in the concentration of gold in the solution. The experiment was carried out for 6 h in distilled water. The results indicate that gold was not released to the solution. Therefore, it can be assumed that the deposition of Au on hexagonal boron nitride is stable (Figure S3B, Supplementary materials).

Zeta potential was measured for h-BN and for nanocomposite. The particles with values of zeta potential between +20 and −20 mV are considered rather unstable. Particles with values more positive than +20 mV and more negative than −20 mV are normally considered stable [37]. Both the nanocomposite and the pristine material showed a negative value of zeta potential, with values of −27.6 ± 5.95 mV, and −29.7 ± 6.81, respectively.

3.2. In Vitro Microscopic Analyses

Microscopic observations were carried out after 12-h incubation to record the effect of the h-BN_AuNP nanocomposite on both cell lines. As shown in Figure 7, the cells observed under a light microscope exposed to the nanocomposite at a concentration of 10.0 μg mL^{-1} exhibited visible differences in comparison to the control culture (Figure S4A,C, Supplementary materials). In the case of L929 cell line, h-BN_AuNP affected the cellular membrane. Numerous small membrane vesicles were observed in L929 cells. The shape of L929 cells was not changed and the cells showed no tendency for detachment (Figure 7A). In contrast, MCF-7 cells did not exhibit evident changes in the shape and membrane vesicles (Figure 7B). The effect of h-BN_AuNP nanocomposite in MFC-7 cells could be observed by the presence of vacuoles in the cell cytoplasm. Lower cell count was also found in the experimental MCF-7 culture in comparison to the MCF-7 control culture (Figure 7E).

Similarly, time-lapse image sequences were taken using a HoloMonitor™ M4. L929 and MCF-7 cell lines were exposed to the h-BN_Au nanocomposite at a concentration of 10.0 µg mL^{-1} for 24, 48 and 72 h (Figure 7). L929 cells did not show any significant differences in the presence of the nanocomposite (Figure 7B–D). The DT value was determined basing on image sequences. The DT value for the L929 control was 16.82 h, while the DT values of the experimental culture were 17.16 h, 22.29 h and 22.47 h for 24-, 48- and 72-h incubation, respectively. L929 cells showed no reduction in proliferation under experimental conditions for cultures exposed to h-BN_AuNP during a 24-h incubation, whereas cell doubling times for 48- and 72-h incubations were higher in comparison to the control culture.

Results obtained for the MCF-7 cell line incubated with h-BN_AuNP demonstrated a stronger effect on the cells at a concentration of 10.0 µg mL^{-1} (Figure 7F–H). MCF-7 cells did not show any visible morphological changes in comparison to the control culture, but the DT analysis indicated a reduction in proliferation capacity. The DT value for the MCF-7 control sample was 42.41 h, whereas the doubling time for experimental cultures was 52.93 h, 68.70 h, and 97.89 h for 24-, 48-, and 72-h incubations, respectively.

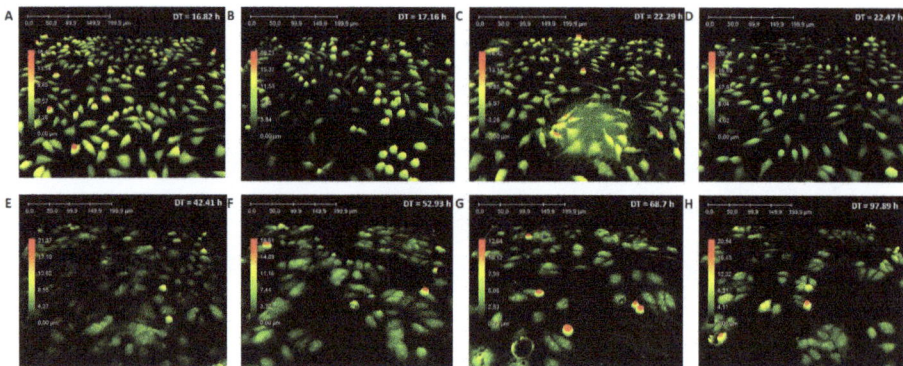

Figure 7. Morphology of the L929 and the MCF-7 cell lines incubated with the h-BN_AuNP nanocomposite. L929 control culture (**A**), L929 culture at 24 h (**B**), 48 h (**C**) and 72 h (**D**), MCF-7 control culture (**E**), MCF-7 culture at 24 h (**F**), 48 h (**G**) and 72 h (**H**) (DT—doubling time).

3.3. Analysis of Cytotoxicity Results

Cytocompatibility of the h-BN_Au nanocomposite at 3.125, 6.25, 10.0, 12.5, 25.0, 50.0, 100.0 and 200.0 µg mL^{-1} concentrations was determined using CCK-8, LDH and NRU assays (Figure 8A–F). Both selected cell lines, L929 and MCF-7, exhibited minimal reduction in mitochondrial activity in the CCK-8 assay. The highest reduction of the mitochondrial metabolism was recorded at the concentration of 200.0 µg mL^{-1} for L929 cells incubated for 48 and 72 h (Figure 8C,E). Mitochondrial activity at the concentration of 10.0 µg mL^{-1} and 200.0 µg mL^{-1} was reduced to 75% and 60%, respectively, compared to free-grown L929 cells. In contrast, mitochondrial activity of MCF-7 cells decreased to 80% at a concentration of 200.0 µg mL^{-1} during a 72-h incubation compared to the control cultures (Figure 8F).

The integrity of L929 cell plasma membranes was affected the most by the h-BN_Au nanocomposite in the range of 100.0–200.0 µg mL^{-1} (Figure 8B,F). LDH leakage at 100.0 µg mL^{-1} was increased for L929 cells by 8% in comparison to control samples and by 15% compared to control samples at the 200.0 µg mL^{-1} concentration during the 24-h exposure. L929 cells incubated with the novel nanocomposite for 72 h at the 200.0 µg mL^{-1} concentration exhibited an increase in LDH leakage by 7% compared to the control samples. As presented in Figure 8A–F, h-BN_AuNP did not affect the integrity of L929 cell membranes in a dose-dependent manner. Lactate dehydrogenase leakage was observed in the MCF-7 human breast adenocarcinoma cell line during the cell incubation with the h-BN_Au nanocomposite within the concentration range of 10.0–200.0 µg mL^{-1} (Figure 8B,D,F) and it did not influence the cells in a dose-dependent manner. The highest LDH release was 38% for the 100.0 µg mL^{-1} concentration of tested nanomaterial after a 72-h incubation. The h-BN_Au nanocomposite, at concentrations ranging from 3.125 to 12.5 µg mL^{-1}, did not cause any disruption of the cell membrane integrity within 24 h. Longer exposure to the nanocomposite resulted in LDH release.

In contrast to LDH assay results, the neural red uptake (NRU) assay showed different tendency with respect to the viability of both cell lines (Figure 8). As regards the L929 cell line, the relative viability decreased to approximately 45–85% at the 3.125–100.0 µg mL^{-1} h-BN_AuNP concentration range (Figure 8A,C,E). The lowest relative viability was observed when L929 fibroblasts were incubated with h-BN_AuNP at concentrations of 10.0, 12.5 and 200.0 µg mL^{-1}. Similarly, the viability of MCF-7 cells was also reduced. The relative viability was reduced to approximately 55–60% in the range of 3.125–25.0 µg mL^{-1} of the nanomaterial, whereas the viability of MCF-7 cells at higher doses of the nanomaterial (50.0–200.0 µg mL^{-1}) decreased to 20–60% compared to control cultures (Figure 8B,D,F).

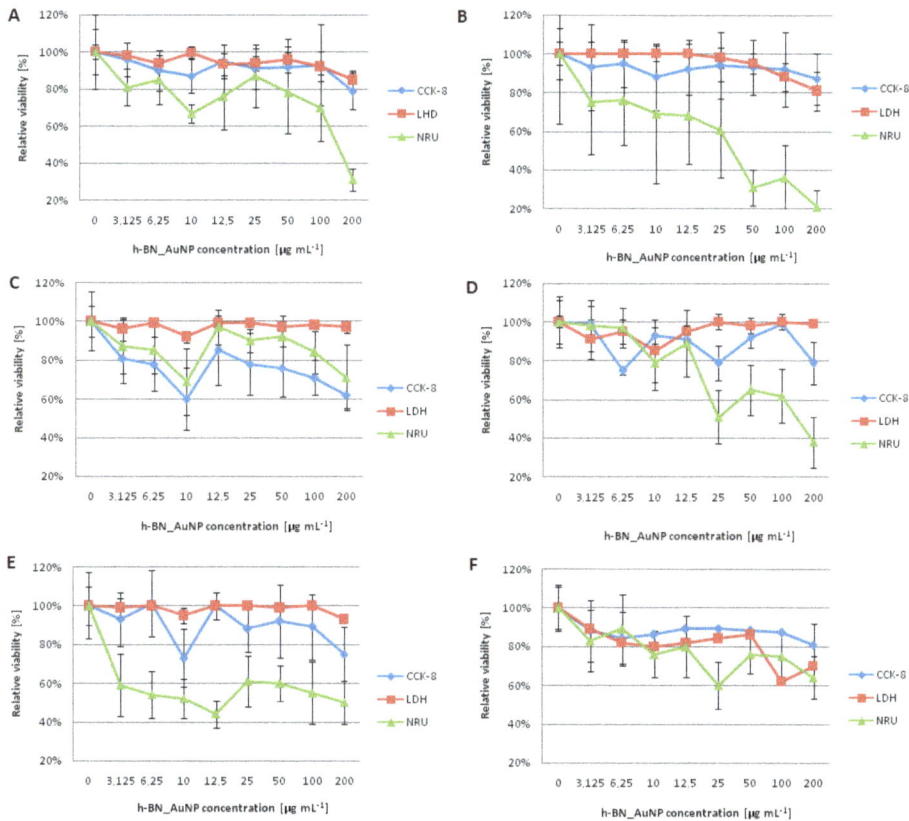

Figure 8. Cytocompatibility analysis based on L929 (24 h—**A**, 48 h—**C**, 72 h—**E**) and MCF-7 cell line (24 h—**B**, 48 h—**D**, 72 h—**F**). Bars represent standard deviation and different small letters ('a'—for CCK-8 results, 'b'—for LDH results, 'c'—for NRU results) indicates statistically significant difference ($p < 0.05$).

3.4. Cellular Uptake Results

The uptake process of hexagonal boron nitride nanoflakes labeled with FITC by normal and cancer cells in three different time points (24, 48 and 72 h) was visualized using confocal microscopy. Figure 9 presenting the intercellular localization of h-BN-FITC in cell cytoplasm—it was confirmed by green fluorescent signal (generated by FITC used for labeling h-BN structures, FT-IR spectrum for the h-BN_FITC was presented in Figure S5, Supplementary materials). The abundance of h-BN was accumulated in the perinuclear region. The presence of h-BN was not confirmed in the nucleus.

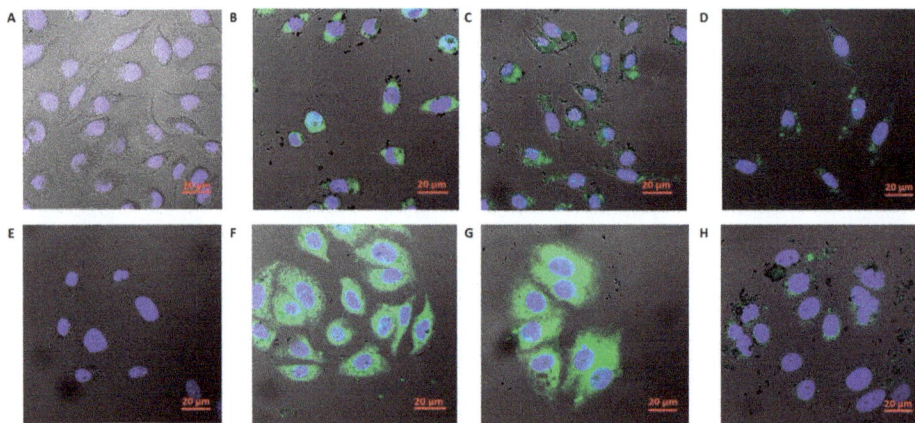

Figure 9. Confocal laser scanning microscopy images. L929 and MCF-7 cells incubated with h-BN labeled with FITC at concentration of 50.0 μg mL^{-1}. L929 control culture (**A**), L929 culture at 24 h (**B**), 48 h (**C**) and 72 h (**D**), MCF-7 control culture (**E**), MCF-7 culture at 24 h (**F**), 48 h (**G**) and 72 h (**H**).

4. Discussion

4.1. The Cytotoxicity of Hexagonal Boron Nitride

The hexagonal boron nitride (h-BN) compound has a high chemical stability and its colorlessness makes this material convenient for detection of the optical signals derived from DNA, proteins and/or other molecules. On the other hand, h-BN is hardly soluble due to inertness of BN structure, which limits its integrations into biological systems. Thus, many approaches have been developed to solve the problem of BN materials' solubility (e.g., by functionalization of BN surface with hydroxyl groups, alkyl chains, interactions between BN with guest molecules) [21]. The concentrations for the stabilized BNs are limited. As it was stated by Zhi and co-workes (2005) and Wang et al. (2008), typically, BNs are stable at concentrations lower than the order of 0.01 mg mL^{-1} [9,38,39]. For the same reason, the analysis of h-BN impact on living matter (in vitro and in vivo) is undoubtedly important [40]. Lu et al. [40] reported that highly dispersed water-soluble ultrathin h-BN nanoplates (of approx. 30–60 nm in diameter and 16 nm in thickness) did not increase the apoptotic rate of HEK-293T (human embryonic kidney cells 293) and CHO (Chinese hamster ovary) cells in the apoptosis assay. The classical MTT (2-(4,5-dimethyl-2-thiazolyl)-3,5-diphenyl-2H-tetrazolium bromide) assay also confirmed high biocompatibility of h-BN in the concentration range of 0–100.0 μg mL^{-1} after 48-h incubation [40]. Ultrathin hexagonal boron nitride nanoplates were also prepared and tested by Nurunnabi et al. [41]. The exfoliated h-BN was partially functionalized with -OH groups in the latter study and h-BN-OH was analyzed to evaluate the potential cytotoxicity by incubating KB cells with h-BN-OH (at concentrations of 10, 20, 50, 100, 250 and 500 μg mL^{-1}) for 24 h. Relative viability of KB cells was measured using the MTT assay. The cytotoxicity of h-BN-OH was not significant even at the highest concentration and cell viability was higher than 90%. In addition, the hemolysis assay demonstrated that h-BN-OH is a blood-compatible material (hemolysis was approximately 4% at the 100 µg mL^{-1} concentration) [41].

Weng and co-workers (2014) developed the solid reaction resulting in highly water-soluble and porous BNs. Via thermal substitution of C atoms with boric acid (H$_3$BO$_3$) substructures in graphitic carbon nitrides (g-C$_3$N$_4$), the hydroxylated BN structure with 60% B atoms with hydroxyl groups was obtained. The BN(OH)$_x$ (x = 0.6–0.9) displays high hydroxylation degrees and forms stable water solution (2.0 mg mL^{-1}) [21]. Novel highly water solubility BNs demonstrated low cytotoxicity (CCK-8 assays showed that more than 92% cells displayed viability after 24 h exposure to 100 µg mL^{-1}) and were tested to determine drugs delivery and releasing capability. The human prostate cancerous cells (LNCaP) exhibited dose-dependent sensitivity to the DOX@BN(OH)$_x$ during 24-h exposure. Moreover, the DOX-loaded BN materials exhibited higher cytotoxicity than free doxorubicin [21]. In another study Li et al. (2017) analyzed hollow BN spheres with controlled crystallinity and solubility in two prostate cancer cell cultures—LNCap (androgen-sensitive) and DU-145 (androgen-independent). The fabricated hollow BN spheres were used as a carrier of boric acid (BA) and source of boron (B) that was released by adjusting the temperature. The cells treated with hollow BN spheres and with BN spheres with controlled release of B demonstrated to inhibit the proliferation and apoptosis. caspase-3/7 activity and LDH release confirmed therapeutic property of tested nanomaterials. Furthermore, in vivo analysis in male BALB/c-nu/nu mice demonstrated the suppression of prostate cancer and inhibition of tumor growth [42]. Chan et al. [43] developed binary polypropylene (PP) composites with hexagonal boron nitride and ternary hybrids with h-BN and nanohydroxyapatite (nHA) and tested the obtained materials using osteoblastic cell culture and the MTT assay. Based on the results, Chan et al. [43] concluded that osteoblasts were able to attach to the surface of PP/h-BN, PP/h-BN-nHA and proliferate. The MTT assay showed the cytotoxicity in the rage of 60–75% for 4- and 7-day incubation periods [43].

In the studies based on boron nitride nanotubes (BNNTs) it was stated that the up-take mechanism may be depended on the BNNT coating as was demonstrated by Chen and co-workers [25,44,45]. The use of biopolymers (such as PLL, GC or PD) for coating of the BNNTs affected the cellular internalization process. The coated BNNTs were easily up-taken by several cell lines. The fluorescence labeled BNNTs showed that this type of nanomaterial was uptaken after 6–12 h of incubation. The TEM analysis showed that the BNNTs were predominantly located in membrane vesicles [46,47]. In the experiment of Ciofani and co-workers (2014), BNNTs were dispersed and stabilized in aqueous solution by the addition of Arabic gum and the effect of gum-coated ultra-pure boron nitride was evaluated on SH-SY5Y (human neuroblastoma) and HUVEC (human umbilical vein endothelial) cell lines. The BNNTs were well tolerated by both cell lines at the concentrations up to 20 µg mL^{-1}. The viability, proliferation, ROS production and apoptosis level were determined and confirmed positive interaction of BNNTs with in vitro models [48]. In our study the uptake of h-BN nanoplates was also confirmed by the visualization of h-BN in cells due to conjugation of nanoplates with FITC. It was found that after 24-h incubation periods nanostructures were present in cellular cytoplasm and perinuclear regions. Other studies on the BNNTs and h-BN as nanocarriers for doxorubicin (Dox) with folate used as targeting agent, were carried out. The Dox loading onto the h-BN was threefold lower than the BNNTs, thus the cytotoxicity study focused only on Dox-BNTTs. It was found that the cellular up-take of folate-Dox-BNNTs was significantly higher than Dox-BNNTs for HeLa cells. The cellular cytotoxicity was determined using HeLa and HUVEC cells. During 8, 24 and 72-h incubation the highest cytotoxic effect was obtained for FA-Dox-BNNTs with significant reduction in cell viability (the viability was in the range of 5 to 40% for HeLa cells and in the range of 20 to 60% for HUVEC). The Dox-BNNTs exhibited cytotoxic effect in the range of 20 to 60% (the same effect was obtained using free Dox) [49]. Feng and co-workes [50] used BN nanospheres (BNNS) conjugated with folate as nanocarriers for doxorubicin (Dox). Obtained BNNS-FA/DOX was recognized and effectively internalized by HeLa cells. It was noticed that HeLa cell proliferation was significantly reduced in the presence of DOX-loaded BNNS. Moreover, BNNS-FA/DOX exhibited higher cytotoxicity than free DOX. Feng and co-workes stated that BNNS-FA complexes are effective drug delivery vehicles [50].

Singh and co-workers [51] tested the biocompatibility of boron nitride in the context of Boron Neutron Therapy (BNCT). The nanostructured BN was added to various cell line cultures: HeLa, HEK-293 and MCF-7 in the concentration range of 0.25–1.0 mg mL^{-1} and incubated for 24 and 48 h to determine the cellular response. The authors observed that cancerous cell lines (HeLa and MCF-7) demonstrated lower relative viability in the MTT assay. Cell survival rate was the lowest after 48-h incubation and reached 40% for MCF-7 and 30% for HeLa cells at the 2 mg mL^{-1} concentration. HEK-293 cells exhibited by 50% higher survival rate during 48-h incubation and by 70% during 24-h exposure at 2 mg mL^{-1} concentration, thus Singh et al. [51] concluded that the cytotoxicity of BN nanostructures was higher for cancerous cells in comparison with normal cell lines [50]. Singh et al. [51] suggested that a higher membrane pore size might be responsible for the enhanced uptake of BN, which led to a higher cytotoxicity, but these conclusions were not validated and need to be evaluated in further investigations [51]. These findings partially confirmed the effect obtained in our study. The analysis of the nanomaterial internalization indicated that it proceeded most efficiently in the first 24 h of incubation and appeared to be more intensive in MCF-7 cells. A significant amount of h-BN_AuNP was accumulated in the cytoplasm of the cells of both lines. Effective internalization mainly translated into the cells' ability to uptake neutral red, reducing it to 30% for cells of the L929 line and 20% for the MCF-7 line. Higher toxicity was found for 10 µg mL^{-1} for the h-BN-Au_NPs in L929 and MCF-7 cell cultures. This cytotoxicity results combined with the stability study (in this study the nanocomposite was prepared as a solution in the PBS/Pluronic F-127 and the best stability was demonstrated at a concentration of 12.5 µg mL^{-1} and was considered as stable; Figure 6) one can conclude that the 10 µg mL^{-1} concentration forms the stable solution and affects more effectively the mitochondrial cells activity that 25 or 50 µg mL^{-1}. We can also conclude that the concentrations of 3.125 and 6.25 µg mL^{-1} were too low to reduce L929 cell relative viability and MCF-7 after 24- and 48-h incubation. In addition, the population doubling time changed slightly over 24 h for the L929 line, while it was extended by about 10 h for MCF-7 cells compared to the control culture. The internalization of the nanomaterial occurred less effectively in the next 24 and 48 h, as a result of aggregate formation in the biological environment (cell culture medium abundant in proteins). The dispersed part of h-BN_Au penetrated inside the cells, while the nanomaterial, in the form of aggregates, was accumulated mainly on the cell surface [52]. Chemical nature of h-BN is also crucial due to the limited solubilization in aqueous solutions. The incubation of cells for 48 h did not affect the DT value for the L929 line, while the population doubling time of MCF-7 cells was longer. A similar observation for the cells incubated with the nanocomposite for a period of 72 h occurred. During 72-h incubation, only the NRU test demonstrated negative effect of the nanocomposite on L929 cells, whereas the MCF-7 line cells showed reduced viability in NRU as well as CCK-8 (tetrazolium-8-[2-(2-methoxy-4-nitrophenyl)-3-(4-nitrophenyl)-5-(2,4-disulfophenyl)-2H-tetrazolium] monosodium salt) and LDH tests. The decrease in relative viability of MCF-7 cells may be related to the increased pore size in the cell membrane and the increased internalization of the nanocomposite (Figure 9).

Although the results of CCK-8 and LDH assays did not indicate a significant cytotoxic effect of h-BN_Au NPs during 24-h incubation, the cell response monitored using the NR uptake assay demonstrated an opposite trend. Neutral red dye uptake is a measure of functional lysosomes, unlike CCK-8 and LDH assays that are based on enzymatic activity. It was reported previously by Weyermann et al. [46] that different cytotoxicity assays can give different or even divergent results depending on the analyzed agent (e.g., drugs, nanomaterials) and the assay employed in the experiment [53]. The tested agent can evoke different cellular responses and engage various intracellular mechanisms (e.g., effect on lysosomes without affecting the cell membrane, impact on specific organelles or general cytotoxicity) [53]. The same tendency was noticed by Fotakis and Timbrell [54] in the study on the comparison of LDH, NR, MTT and protein assays. In some cases (e.g., CdCl$_2$ effect on HepG2 and HCT cells), the cytotoxicity could be observed in the NR assay, whereas no cytotoxic effect was found when other assays were employed. These authors also stated that results were dependent on the cell line type, potentially cytotoxic agents, time of exposure and type of assay used in the experiment. In the work of Fotakis and Timbrell, LDH leakage and protein assays were found less sensitive in comparison to NR and MTT assays and considered more suitable in detecting early toxicity [54]. Presented conclusions are consistent with our results of CCK-8, LDH and NR uptake assays. Although CCK-8 and LDH assays showed that h-BN_Au NPs were biocompatible, the Neutral Red uptake assay demonstrated that lysosome function was affected in both cell lines analyzed in three time points.

Results obtained in different experiments in vitro were confirmed in vivo. The in vivo effect of boron nitride nanomaterial was investigated by Ciofani et al., Wang et al., Liu et al. [55–57]. The boron nitride nanotubes (BNNTs) in aqueous solution were stabilized with polymer—glycol chitosan (G-chitosan) at a 1:1 ratio. The prepared solution of BNNTs coated with G-chitosan at the concentration of 1 mg mL^{-1} was injected into the marginal ear vein of rabbits. The physiological parameters were analyzed at intervals of 0, 2, 24 and 72 h. The hematological analyses (e.g., white cell count, red cell count, platelet count, etc.) did not demonstrate significant differences in the experimental and control groups. At 72 h after injection only platelet count was higher in comparison with control group, but recorded values were in the healthy range for rabbits. The biochemical parameters (renal and hepatic) in experimental group were also closely proportional to the values in the control group. This study demonstrated the absence of negative effects not only on blood parameters, but also on liver and kidney functions [48]. In other interesting study, Salvetti et al. [58] tested multiwalled BNNTs on freshwater planarians. The solution of multiwalled BNNTs was prepared with the addition of a 0.1% Arabic gum and the BNNTs coated with Arabic gum (single injection of 100 or 200 µg g^{-1}) were injected into the gut of planarians (*Dugesia japonica*). The short-term effect of BNNTs was determined 4 or 24 h after injection. The long-term (chronic) effect analysis was conducted by the injection of BNNTs twice a week for 15 days (total amount of 100 or 200 µg g^{-1}). The TEM analysis of intestinal cells did not confirmed any physiological changes in BNNT-treated animals with the respect to the control. The nanomaterial was found inside cytoplasmic vesicles of intestinal phagocytes. The oxidative stress and apoptosis level was determined in control and in groups of animals of acute and chronic exposure to nanomaterial. Data obtained by Salvetti et al. [58] indicated that BNNTs did not induce DNA damage and apoptosis (number of apoptotic cells in BNNT-treated animals in comparison with control group). The effect of BNNTs was also analyzed in the context of planarian stem cells and stem cell progeny by means of the real-time PCR method. The expression of marker genes for proliferating neoblasts, early neoblast progeny and late neoblast progeny, was not affected significantly in the group of BNNT-treated animals. Moreover the effect of BNNTs on regeneration process was also monitored. Both acute and chronic BNNT-treated animals did not exhibit morphological abnormalities and the regeneration processes were not compromise [58].

4.2. Presence of Au Nanoparticles on h-BN Nanoplates and Their Impact on Cell Activity and Proliferation Rate

The recent increasing interest in medical applications of gold nanoparticles encouraged scientists to analyze their potential impact on biological systems. Gold nanoparticles were found to be useful in wound healing and infection prevention. Currently, many studies are concentrated around the anticancer properties of gold nanoparticles. Kamala Priya et al. [59] studied GNPs in MCF 7 cell line. The incubation of cells in the presence of GNPs resulted in an efficient cellular metabolism reduction even at the minimum concentration of 2 µg mL^{-1} [59]. The significance of gold nanoparticles in nanobiotechnology was also evaluated by Rattanata et al. [60]. Gold nanoparticles conjugated with gallic acid (GA) (at concentrations of 30–150 µM) were analyzed in vitro using M213 and M214 cell cultures. These authors found that GNPs-GA complexes inhibited cancer cell proliferation and caused changes in cellular membrane lipids and fatty acids, which resulted in cell death via apoptosis [60]. Geetha et al. [61] tested newly synthesized gold nanoparticles for their anti-leukemic cancer activity (HL-60 cells). The analysis was based on the MTT assay, DNA fragmentation, apoptosis and comet assays. The results obtained by Geetha and co-workers showed HL-60 response to gold nanoparticles treatment that involved DNA fragmentation (gel electrophoresis showed enhanced DNA fragmentation with increasing exposure to gold nanoparticles and longer DNA tail present in the comet assay) and reduction of relative cell viability (in the MTT assay) [61]. Gold nanoparticles (GNPs) are also considered to be an effective delivery platform. Fu et al. [62] loaded Dox directly on GNPs by forming hydrazone (HDox-GNPs) or amide (SDox-GNPs) bonds. Their cytotoxicity and anticancer efficacy were evaluated using U87, HeLa, MCF-7 and A549 cell cultures. HDox-GNPs and SDox-GNPs were tested in a series of concentrations (10, 50, 100, and 500 nM; 1, 5, and 10 μM) for 72 h. mGNPs (monofunctional gold nanoparticles) exhibited low cytotoxicity, whereas HDox-GNPs showed evident cytotoxicity (IC50 for U87 cells was 0.19). SDox-GNPs exhibited much lower cytotoxicity against U87 cells. The difference in IC50 for SDox-GNPs was 2- to 18-fold higher compared to HDox-GNPs [62].

From the chemo-physical point of view, the effect of gold nanoparticles may be the result of their unique and well-known catalytic activities. Those structures supported on metal oxides can catalyze various oxidation reactions by molecular oxygen [63,64]. The size of clusters is an important factor affecting the gold activity. The unique catalytic activity emerges when the size of clusters decreases down to 1-5 nm. The catalytic properties of Au and Au_2, supported on the hexagonal boron nitride (h-BN) surface or impurity point defects in h-BN absorption and activation of O_2, can be affected by two different mechanisms. The ability of Au and Au_2 to activate O_2 is dependent on the electron pushing mechanism and donor/acceptor mechanism [63,64]. Gao et al. [63] reported that weak interactions of gold particles with h-BN led to binding the stimulation and catalytic activation of O_2 absorbed on Au/h-BN [63]. Strong absorption of surface defects is related to the charge transfer from/to the adsorbate. That charge transfer can affect the gold catalytic activity. The interaction between Au and the h-BN surface can affect the CO oxidation reaction and oxygen reduction reaction (ORR) by O_2. The defect-free h-BN surface promotes the electron transfer from Au to O_2 (pushing electrons from gold to adsorbed oxygen) [65,66].

From a biological standpoint, transition metal nanoparticle were found to induce the chromosomal aberrations, DNA strand breaks, oxidative DNA damage and mutations. DNA single strand breakage can be induced by $OH^\bullet$ via formation of 8-hydroxylo-2'-deoxyguanosin (8-OHdG) DNA adduct. In vivo and in vitro experiments demonstrated that NPs, including Cu, Fe, Ti and Ag metal oxides cause micronuclei and DNA damage [67]. Oxygen free radicals, such as superoxide ($O_2^{\bullet-}$) and hydroxyl ($OH^\bullet$) radicals and other reactive species (ROS), e.g., hydrogen peroxide (H_2O_2) are mediatiors promoting growth, metabolic or cytostatic effects. Many cellular events are regulated by changes in the redox status [68]. It is known that oxidative stress is responsible for the reduction of the proliferation rate by the inhibition of the transition of cells from the G0 to the G1 phase. As a result, the G1 phase is prolonged, the S phase progresses slower due to DNA synthesis inhibition, cell cycle is inhibited through the restriction point and eventually arrested at the cell cycle checkpoints. Some studies showed that the proliferation rate of normal and cancer cells were decreased in the periods of oxidative stress. Other experiments showed that only tumor cell growth was inhibited in the cell cultures as well as in laboratory animals. The oxidative stress affects the rate of cell proliferation by the inhibition of crucial enzymes (e.g., DNA polymerases and cyclin-dependent kinase) [69].

The response of tumor cells to chronic oxidative stress generated in radiotherapy, photodynamic therapy as well as many chemotherapies, is used in anticancer therapies. Antitumor activity is a relationship between the induction of apoptosis and DNA damage induced by oxygen radicals. Cancer cells may be more sensitive to ROS accumulation in comparison to normal cells [70]. Increased production of ROS that reaches the threshold (incompatible with cell viability) enhances the antioxidant mechanism leading to selective cancer cell death without affecting normal cells [71]. On the other hand, one must take into consideration that the oxidative stress within tumor cells may cause resistance to therapy by increasing cellular expression of P-glycoprotein [72].

Nanomaterials **2018**, *8*, 605

5. Conclusions

In our preliminary study, hexagonal boron nitride was exfoliated using chemical and physical methods and it was functionalized to obtain nanohybrid of h-BN with Au particles. The nanomaterial was characterized to determine the sample quality. The cytotoxicity of h-BN_Au particles was evaluated and it was found that h-BN_AuNPs did not affect the cellular metabolism (CCK-8 and LDH assays), but it had an impact on the function of lysosomes in both normal and cancer cell lines during 24-h exposition. Longer incubation for 48- and 72-h affected the cell relative viability at the concentration of 10 µg mL^{-1}. Moreover, h-BN_Au particles demonstrated inhibition of proliferative activity of the MCF-7 cancer cell line in comparison with normal L929 cell line after 72-h incubation period. These results make the new hybrid nanomaterial an interesting tool not only for anticancer therapy, but also can be used as a platform in biosensor design or in tissue engineering. It is worth emphasizing that the effect of the h-BN_Au nanoparticle on living structures should be examined in greater detail.

Supplementary Materials: The following are available online at http://www.mdpi.com/2079-4991/8/8/605/s1.

Author Contributions: M.J.-S. and E.M. conceived and designed the experiments; M.J.-S., M.T. and M.D. performed the experiments; M.J.-S., M.T. and K.P. analyzed the data; E.M. contributed reagents/materials/analysis tools; M.J.-S. and M.D. wrote the paper.

Funding: This research was funded by National Science Centre (Poland) within the project No. 2016/21/N/ST8/02397 (PRELUDIUM 11).

Acknowledgments: The authors are grateful to Pawel Sznigir for assistance in performing experiments and providing technical support (HM4 HoloMonitor™).

Conflicts of Interest: The authors declare no conflict of interest.

References

1. Ferrari, A.C.; Bonaccorso, F.; Fal'ko, V.; Novoselov, K.S.; Roche, S.; Bøggild, P.; Borini, S.; Koppens, F.H.; Palermo, V.; Pugno, N.; et al. Science and technology roadmap for graphene, related two-dimensional crystals, and hybrid systems. *Nanoscale* **2015**, *7*, 4598–4810. [CrossRef] [PubMed]
2. Weng, Q.; Wang, X.; Wang, X.; Bando, Y.; Golberg, D. Functionalized hexagonal boron nitride nanomaterials: Emerging properties and applications. *Chem. Soc. Rev.* **2016**, *45*, 3989–4012. [CrossRef] [PubMed]
3. Dobrzhinetskaya, L.F.; Wirth, R.; Yang, J.; Green, H.W.; Hutcheon, I.D.; Weber, P.K.; Grew, E.S. Qingsongite, natural cubic boron mineral from the Earth's mantle. *Am. Miner.* **2014**, *99*, 764–772. [CrossRef]
4. Golberg, D.; Bando, Y.; Huang, Y.; Terano, T.; Mitome, M.; Tang, C.; Zhi, C. Boron nitride nanotubes and nanosheets. *ACS Nano* **2010**, *4*, 2979–2993. [CrossRef] [PubMed]
5. *Safety Assessment of Boron Nitride as Used in Cosmetics*; Scientific Literature Review for Public Comment; Cosmetics Ingredient Review: Washington, DC, USA, 2012.
6. Sediri, H.; Pierucci, D.; Hajlaoui, M.; Henck, H.; Patriarche, G.; Dappe, Y.J.; Yuan, S.; Toury, B.; Belkhou, R.; Silly, M.G.; et al. Atomically sharp interface in an h-BN-epitaxial graphene van der Waals heterostructure. *Sci. Rep.* **2015**, *5*, 16465. [CrossRef] [PubMed]
7. Wood, G.L.; Paine, R.T. Aerosol synthesis of hollow spherical morphology boron nitride particles. *Chem. Mater.* **2006**, *18*, 4716–4718. [CrossRef]
8. Cai, Q.; Mateti, S.; Yang, W.; Jones, R.; Watanabe, K.; Taniguchi, T.; Huang, S.; Chen, Y.; Li, L.H. Boron nitride nanosheets improve sensitivity and reusability of surface enhanced Raman spectroscopy. *Angew. Chem. Int. Ed.* **2016**, *55*, 8405–8409. [CrossRef] [PubMed]
9. Wang, J.; Ma, F.; Liang, W.; Sun, M. Electrical properties and applications of graphene, hexagonal boron nitride (h-BN), and graphene/h-BN heterostructures. *Mater. Today Phys.* **2017**, *2*, 6–34. [CrossRef]
10. Falin, A.; Cai, Q.; Santos, E.J.G.; Scullion, D.; Qian, D.; Zhang, R.; Yang, Z.; Huang, S.; Watanabe, K.; Taniguchi, T.; et al. Mechanical properties of atomically thin boron nitride and the role of interlayer interactions. *Nat. Commun.* **2017**, *8*, 15815. [CrossRef] [PubMed]

11. Qi, H.; Wang, L.; Sun, J.; Long, Y.; Hu, P.; Liu, F.; He, X. Production methods of van der Waals heterostructures based on transition metal dichalcogenides. *Crystals* **2018**, *8*, 35. [CrossRef]

12. Jariwala, D.; Marks, T.J.; Hersam, M. Mixed-dimensional van der Waals heterostructures. *Nat. Mater.* **2017**, *16*, 170. [CrossRef] [PubMed]

13. Zhao, S.F.; Elbaz, G.A.; Bediako, D.K.; Yu, C.; Efetov, D.K.; Guo, Y.; Rivichandran, J.; Min, K.A.; Hong, S.; Taniguchi, T.; et al. Controlled electrochemical intercalation of graphene/h-BN van der Waals heterostructures. *Nano Lett.* **2018**, *18*, 460–466. [CrossRef] [PubMed]

14. Lee, C.-H.; Schiros, T.; Santos, E.J.G.; Kim, B.; Yager, K.G.; Kang, S.J.; Lee, S.; Yu, J.; Watanabe, K.; Taniguchi, T.; et al. Epitaxial growth of molecular crystals on van der Waals substrates for high-performance organic electronics. *Adv. Mater.* **2014**, *26*, 2812–2817. [CrossRef] [PubMed]

15. Lee, G.-H.; Lee, C.-H.; van der Zande, A.M.; Han, M.; Cui, X.; Arefe, G.; Nuckolls, C.; Heinz, T.F.; Hone, J.; Kim, P. Heterostructures based on inorganic and organic van der Waals systems. *APL Mater.* **2014**, *2*, 092511. [CrossRef]

16. Jariwala, D.; Sangwan, V.K.; Lauhon, L.J.; Marks, T.J.; Hersam, M.C. Carbon nanomaterials for electronics, optoelectronics, photovoltaics, and sensing. *Chem. Soc. Rev.* **2013**, *42*, 2824–2860. [CrossRef] [PubMed]

17. Kobayashi, Y.; Kumakura, K.; Akasaka, T.; Makimoto, T. Layered boron nitride as a release layer for mechanical transfer of GaN-based devices. *Nature* **2012**, *484*, 223–227. [CrossRef] [PubMed]

18. Chung, K.; Lee, C.-H.; Yi, G.-C. Transferable GaN layers grown on ZnO-coated graphene layers for optoelectronic devices. *Science* **2010**, *330*, 655–657. [CrossRef] [PubMed]

19. Jariwala, D.; Sangwan, V.K.; Lauhon, L.J.; Marks, T.J.; Hersam, M.C. Emerging device applications for semiconducting two-dimensional transition metal dichalcogenides. *ACS Nano* **2014**, *8*, 1102–1120. [CrossRef] [PubMed]

20. Li, M.Y.; Chen, C.H.; Shi, Y.; Li, L.J. Heterostructures based on two-dimensional layered materials and their potential applications. *Mater. Today* **2016**, *19*, 322–335. [CrossRef]

21. Weng, Q.; Wang, B.; Wang, X.; Hanagata, N.; Li, X.; Liu, D.; Wang, X.; Jiang, X.; Bando, Y.; Golberg, D. Highly water-soluble, porous, and biocompatible boron nitrides for anticancer drug delivery. *ACS Nano* **2014**, *8*, 6123–6130. [CrossRef] [PubMed]

22. Gao, Z.; Zhi, C.; Bando, Y.; Golberg, D.; Serizawa, T. Noncovalent functionalization of boron nitride nanotubes in aqueous media opens application roads in nanomedicine. *Nanomedicine* **2014**, *1*, 7.

23. Gottschalck, T.E.; Breslawec, H. *International Cosmetic Ingredient Dictionary and Handbook*; Personal Care Products Council: Washington, DC, USA, 2012.

24. Rasel, A.I.; Li, T.; Nguyen, T.D.; Singh, S.; Zhou, Y.; Xiao, Y.; Gu, Y.T. Biophysical response of living cells to boron nitride nanoparticles: Uptake mechanism and bio-mechanical characterization. *J. Nanopart. Res.* **2015**, *17*, 441. [CrossRef]

25. Chen, X.; Wu, P.; Rousseas, M.; Okawa, D.; Gartner, Z.; Zettl, A.; Bertozzi, C.R. Boron nitride nanotubes are noncytotoxic and can be functionalized for interaction with proteins and cells. *J. Am. Chem. Soc.* **2009**, *131*, 890–891. [CrossRef] [PubMed]

26. Horváth, L.; Magrez, A.; Golberg, D.; Zhi, C.; Bando, Y.; Smajda, R.; Horváth, E.; Forró, L.; Schwaller, B. In vitro investigation of the cellular toxicity of boron nitride nanotubes. *ACS Nano* **2011**, *5*, 3800–3810. [CrossRef] [PubMed]

27. Khan, A.F.; Randviir, E.P.; Brownson, D.A.; Ji, X.; Smith, G.C.; Banks, C.E. 2D hexagonal boron nitride (2D-hBN) explored as a potential electrocatalyst for the oxygen reduction reaction. *Electrocatalysis* **2017**, *29*, 622–634. [CrossRef]

28. Xu, X.; Hao, H.; Khanaki, A.; Zheng, R.; Suja, M.; Liu, J. Large-area growth of multi-layer hexagonal boron nitride on polished cobalt foils by plasma-assisted molecular beam epitaxy. *Sci. Rep.* **2017**, *7*, 43100. [CrossRef] [PubMed]

29. Sudeep, P.M.; Vinod, S.; Ozden, S.; Sruthi, R.; Kukovecz, A.; Konya, Z.; Vajtai, R.; Anantharaman, M.R.; Ajayan, P.M.; Narayanan, T.N. Functionalized boron nitride porous solids. *RSC Adv.* **2015**, *5*, 93964–93968. [CrossRef]

30. Meunier, F.C. In situ FT-IR spectroscopy investigations of dimethyl carbonate synthesis: On the contribution of gas-phase species. *RSC Adv.* **2016**, *6*, 17288–17289. [CrossRef]

31. Esumi, K.; Hosoya, T.; Suzuki, A.; Torigoe, K. Spontaneous formation of gold nanoparticles in aqueous solution of sugar-persubstituted poly(amidoamine)dendrimers. *Langmuir* **2000**, *16*, 2978–2980. [CrossRef]

32. Korsaks, V. Hexagonal boron nitride luminescence dependent on vacuum level and surrounding gases. *Mater. Res. Bull.* **2015**, *70*, 976–979. [CrossRef]
33. Vijayan, S.R.; Santhiyagu, P.; Singamuthu, M.; Ahila, N.K.; Jayaraman, R.; Ethiraj, K. Synthesis and characterization of silver and gold nanoparticles using aqueous extract of seaweed, *Turbinaria conoides*, and their antimicrofouling activity. *Sci. World J.* **2014**, *2014*, 938272. [CrossRef] [PubMed]
34. Cai, Q.; Du, A.; Gao, G.; Mateti, S.; Cowie, B.C.; Qian, D.; Zhang, S.; Lu, Y.; Fu, L.; Taniguchi, T.; et al. Molecule-induced conformational change in boron nitride nanosheets with enhanced surface adsorption. *Adv. Funct. Mater.* **2016**, *26*, 8202–8210. [CrossRef]
35. Griffin, A.; Harvey, A.; Cunningham, B.; Scullion, D.; Tian, T.; Shih, C.J.; Gruening, M.; Donegan, J.F.; Santos, E.J.; Backes, C.; et al. Spectroscopic size and thickness metrics for liquid-exfoliated h-BN. *Chem. Mater.* **2018**, *30*, 1998–2005. [CrossRef]
36. Draz, M.S.; Lu, X. Development of a loop mediated isothermal amplification (LAMP)—Surface Enhanced Raman spectroscopy (SERS) assay for the detection of *Salmonella enterica* serotype *Enteritidis*. *Theranostics* **2016**, *6*, 522. [CrossRef] [PubMed]
37. Ferreira, T.H.; Hollanda, L.M.; Lancellotti, M.; de Sous, E.M.B. Boron nitride nanotubes chemically functionalized with glycol chitosan for gene transfection in eukaryotic cell lines. *J. Biomed. Mater. Res. A* **2015**, *103*, 2176–2185. [CrossRef] [PubMed]
38. Zhi, C.Y.; Bando, Y.; Tang, C.C.; Xie, R.G.; Sekiguchi, T.; Golberg, D. Perfectly dissolved boron nitride nanotubes due to polymer wrapping. *J. Am. Chem. Soc.* **2005**, *127*, 15996–15997. [CrossRef] [PubMed]
39. Wang, W.L.; Bando, Y.; Zhi, C.Y.; Fu, W.Y.; Wang, E.G.; Golberg, D. Aqueous noncovalent functionalization and controlled near-surface carbon doping of multiwalled boron nitride nanotubes. *J. Am. Chem. Soc.* **2008**, *130*, 8144–8145. [CrossRef] [PubMed]
40. Lu, T.; Wang, L.; Jiang, Y.; Iiu, Q.; Huang, C. Hexagonal boron nitride nanoplates as emerging biological nanovectors and their potential applications in biomedicine. *J. Mater. Chem. B* **2016**, *4*, 6103–6110. [CrossRef]
41. Nurunnabi, M.; Nafiujjaman, M.; Lee, S.J.; Park, I.K.; Huh, K.M.; Lee, Y. Preparation of ultra-thin hexagonal boron nitride nanoplates for cancer cell imaging and neurotransmitter sensing. *Chem. Commun.* **2016**, *52*, 6146–6149. [CrossRef] [PubMed]
42. Li, X.; Wang, X.; Zhang, J.; Hanagata, N.; Wang, X.; Weng, Q.; Ito, A.; Bando, Y.; Golberg, D. Hollow boron nitride nanospheres as boron reservoir for prostate cancer treatment. *Nat. Commun.* **2017**, *8*, 13936. [CrossRef] [PubMed]
43. Chan, K.W.; Wong, H.M.; Yeung, K.W.K.; Tjong, S.C. Polypropylene biocomposites with boron nitride and nano-hydroxyapatite reinforcements. *Materials* **2015**, *8*, 992–1008. [CrossRef] [PubMed]
44. Fernandez-Yague, M.A.; Larranaga, A.; Gladkovskaya, O.; Stanley, A.; Tadayyon, G.; Guo, Y.; Sarasua, J.R.; Tofail, S.A.M.; Zeugolis, D.I.; Pandit, A.; et al. Effects of polydopamine functionalization on boron nitride nanotube dispersion and cytocompatibility. *Bioconjug. Chem.* **2015**, *26*, 2025–2037. [CrossRef] [PubMed]
45. Ciofani, G.; Danti, S.; D'Alessandro, D.; Moscato, A.S.; Menciassi, A. Assessing cytotoxicity of boron nitride nanotubes: Interference with the MTT assay. *Biochem. Biophys. Res. Commun.* **2010**, *394*, 405–411. [CrossRef] [PubMed]
46. Ciofani, G.; Danti, S.; Genchi, G.G.; Mazzolai, B.; Mattoli, V. Boron nitride nanotubes: Biocompatibility and potential spill-over in nanomedicine. *Small* **2013**, *9*, 1672–1685. [CrossRef] [PubMed]
47. Ciofani, G.; Ricotti, L.; Danti, S.; Moscato, S.; Nesti, C.; D'Alessandro, D.; Dinucci, D.; Chiellini, F.; Pietrabissa, A.; Petrini, M.; et al. Investigation of interactions between poly-L-lysine-coated boron nitride nanotubes and C2C12 cells: Up-take, cytocompatibility, and differentiation. *Int. J. Nanomed.* **2010**, *5*, 285. [CrossRef]
48. Ciofani, G.; del Turco, S.; Rocca, A.; de Vito, G.; Cappello, V.; Yamaguchi, M.; Li, X.; Mazzolai, B.; Basta, G.; Gemmi, M.; et al. Cytocompatibility evaluation of gum Arabic-coated ultra-pure boron nitride nanotubes on human cells. *Nanomedicine* **2014**, *9*, 773–788. [CrossRef] [PubMed]
49. Emanet, M.; Şen, Ö.; Çulha, M. Evaluation of born nitride nanotubes and hexagonal boron nitride as nanocarriers for cancer drugs. *Nanomedicine* **2017**, *12*, 797–810. [CrossRef] [PubMed]
50. Feng, S.; Zhang, H.; Yan, T.; Huang, D.; Zhi, C.; Nakanishi, H.; Gao, X.D. Folate-conjugated boron nitride nanospheres for targeted delivery of anticancer drugs. *Int. J. Nanomed.* **2016**, *11*, 4573.

51. Singh, B.; Kaur, G.; Singh, P.; Singh, K.; Kumar, B.; Vij, A.; Kumar, M.; Bala, R.; Meena, R.; Singh, A.; et al. Nanostructured boron nitride with high water dispersibility for boron neutron capture therapy. *Sci. Rep.* **2016**, *6*, 35535. [CrossRef] [PubMed]

52. Mateti, S.; Wong, C.S.; Liu, Z.; Yang, W.; Li, Y.; Li, L.H.; Chen, Y. Biocompatibility of boron nitride nanosheets. *Nano Res.* **2018**, *11*, 334–342. [CrossRef]

53. Weyermann, J.; Lochmann, D.; Zimmer, A. A practical note on the use of cytotoxicity assays. *Int. J. Pharm.* **2005**, *288*, 369–376. [CrossRef] [PubMed]

54. Fotakis, G.; Timbrell, J.A. In vitro cytotoxicity assays: Comparison of LDH, neutral red, MTT and protein assay in hepatoma cell lines following exposure to cadmium chloride. *Toxicol. Lett.* **2006**, *160*, 171–177. [CrossRef] [PubMed]

55. Ciofani, G.; Danti, S.; Genchi, G.G.; D'Alessandro, D.; Odorico, M.; Mattoli, V.; Giorgi, M. Pilot in vivo toxicological investigation of boron nitride nanotubes. *Int. J. Nanomed.* **2012**, *7*, 19. [CrossRef] [PubMed]

56. Wang, N.; Wang, H.; Tang, C.; Lei, S.; Shen, W.; Wang, C.; Wang, G.; Wang, Z.; Wang, L. Toxicity evaluation of boron nitride nanospheres and water-soluble boron nitride in *Caenorhabditis elegans*. *Int. J. Nanomed.* **2017**, *12*, 5941. [CrossRef] [PubMed]

57. Liu, B.; Qi, W.; Tian, L.; Li, Z.; Miao, G.; An, W.; Liu, D.; Lin, J.; Zhang, X.; Wu, W. In vivo biodistribution and toxicity of highly soluble PEG-coated boron nitride in mice. *Nanoscale Res. Lett.* **2015**, *10*, 478. [CrossRef] [PubMed]

58. Salvetti, A.; Rossi, L.; Iacopetti, P.; Li, X.; Nitti, S.; Pellegrino, T.; Mattoli, V.; Golberg, D.; Ciofani, G. In vivo biocompatibility of boron nitride nanotubes: Effects on stem cell biology and tissue regeneration in planarians. *Nanomedicine* **2015**, *10*, 1911–1922. [CrossRef] [PubMed]

59. Priya, M.R.K.; Iyer, P.R. Applications of the green synthesized gold nanoparticles-antimicrobial activity, water purification system and drug delivery system. *Nanosci. Technol.* **2015**, *2*, 1–4. [CrossRef]

60. Rattanata, N.; Daduang, S.; Wongwattanakul, M.; Leelayuwat, C.; Limpaiboon, T.; Lekphrom, R.; Sandee, A.; Boonsiri, P.; Chio-Srichan, S.; Daduang, J. Gold nanoparticles enhance the anticancer activity of gallic acid against cholangiocarcinoma cell lines. *Asian Pac. J. Cancer Prev.* **2015**, *16*, 7143–7147. [CrossRef] [PubMed]

61. Geetha, R.; Ashokkumar, T.; Tamilselvan, S.; Govindaraju, K.; Sadiq, M.; Singaravelu, G. Green synthesis of gold nanoparticles and their anticancer activity. *Cancer Nano* **2013**, *4*, 91–98. [CrossRef] [PubMed]

62. Fu, Y.; Feng, Q.; Chen, Y.; Shen, Y.; Su, Q.; Zhang, Y.; Zhou, X.; Cheng, Y. Comparison of two approaches for the attachment of a drug to gold nanoparticles and their anticancer activities. *Mol. Pharm.* **2016**, *13*, 3308–3317. [CrossRef] [PubMed]

63. Gao, M.; Lyalin, A.; Taketsugu, T. Catalytic activity of Au and Au_2 on the h-BN surface: Adsorption and activation of O_2. *J. Phys. Chem. C* **2012**, *116*, 9054–9062. [CrossRef]

64. Lyalin, A.; Nakayama, A.; Uosaki, K.; Taketsugu, T. Functionalization of monolayer h-BN by a metal support for the oxygen reduction reaction. *J. Phys. Chem. C* **2013**, *117*, 21359–21370. [CrossRef]

65. Lyalin, A.; Nakayama, A.; Uosaki, K.; Taketsugu, T. Adsorption and catalytic activation of the molecular oxygen on the metal supported h-BN. *Top. Catal.* **2014**, *57*, 1032–1041. [CrossRef]

66. Gao, M.; Adachi, M.; Lyalin, A.; Taketsugu, T. Long range functionalization of h BN monolayer by carbon doping. *J. Phys. Chem. C* **2016**, *120*, 15993–16001. [CrossRef]

67. Manke, A.; Wang, L.; Rojanasakul, Y. Mechanisms of nanoparticle-induced oxidative stress and toxicity. *BioMed Res. Int.* **2013**, *2013*, 942916. [CrossRef] [PubMed]

68. Halliwell, B. Oxidative stress in cell culture: An under-appreciated problem? *FEBS Lett.* **2003**, *540*, 3–6. [CrossRef]

69. Conklin, K.A. Chemotherapy-associated oxidative stress: Impact on chemotherapeutic effectiveness. *Integr. Cancer Ther.* **2004**, *3*, 294–300. [CrossRef] [PubMed]

70. Liu, J.; Wang, Z. Increased oxidative stress as a selective anticancer therapy. *Oxid. Med. Cell. Longev.* **2015**, *2015*, 294303. [CrossRef] [PubMed]

71. Brown, N.S.; Bicknell, R. Hypoxia and oxidative stress in breast cancer. Oxidative stress: Its effects on the growth, metastatic potential and response to therapy of breast cancer. *Breast Cancer Res.* **2001**, *3*, 323. [CrossRef] [PubMed]
72. Gorbachev, R.V.; Riaz, I.; Nair, R.R.; Jalil, R.; Britnell, L.; Belle, B.D.; Hill, E.W.; Novoselov, K.S.; Watanabe, K.; Taniguchi, T.; et al. Hunting for monolayer boron nitride: Optical and Raman signatures. *Small* **2011**, *7*, 465–468. [CrossRef] [PubMed]

MDPI

St. Alban-Anlage 66

4052 Basel

Switzerland

Tel. +41 61 683 77 34

Fax +41 61 302 89 18

www.mdpi.com

Nanomaterials Editorial Office

E-mail: nanomaterials@mdpi.com

www.mdpi.com/journal/nanomaterials

www.ingramcontent.com/pod-product-compliance
Lightning Source LLC
Chambersburg PA
CBHW051915210326
41597CB00033B/6154